Anonymous

Illustrated Medicine and Surgery

Vol. 1

Anonymous

Illustrated Medicine and Surgery
Vol. 1

ISBN/EAN: 9783337778262

Printed in Europe, USA, Canada, Australia, Japan

Cover: Foto ©berggeist007 / pixelio.de

More available books at **www.hansebooks.com**

ILLUSTRATED

MEDICINE AND SURGERY.

EDITED BY

GEORGE HENRY FOX,
CLINICAL PROFESSOR OF DISEASES OF THE SKIN, COLLEGE OF PHYSICIANS AND SURGEONS, NEW YORK.

AND

FREDERIC R. STURGIS,
PROFESSOR OF VENEREAL DISEASES, MEDICAL DEPARTMENT UNIVERSITY OF THE CITY OF NEW YORK.

WITH THE CO-OPERATION OF

PROFESSORS WILLARD PARKER, A. C. POST, W. H. VAN BUREN, JAS. R. WOOD,
J. L. LITTLE, T. G. THOMAS, A. L. LOOMIS, F. DELAFIELD,
D. B. ST. J. ROOSA, C. R. AGNEW, AND AUSTIN FLINT.

QUARTERLY.

VOLUME I.

NEW YORK:

E. B. TREAT, No. 757 BROADWAY.

ILLUSTRATED

MEDICINE SURGERY

1882.

CONTENTS.

[5]

CONTENTS.

PREFACE.

WITH the close of the first volume the Editors of the "Illustrated Quarterly of Medicine and Surgery" would thank the profession for their kindly support of a new and difficult undertaking.

The aim of the "Quarterly" has been to furnish a medium for the publication of interesting and instructive matter, which would be of little value without suitable illustrations, and which has heretofore been withheld from the profession or published only at personal expense.

The flattering notices of the Medical Press, and a large and increasing subscription list would seem to indicate that a measure of success has already been achieved, and with the continued support of eminent medical men a still greater success is hoped for by the Editors.

GEORGE HENRY FOX.

FREDERIC R. STURGIS.

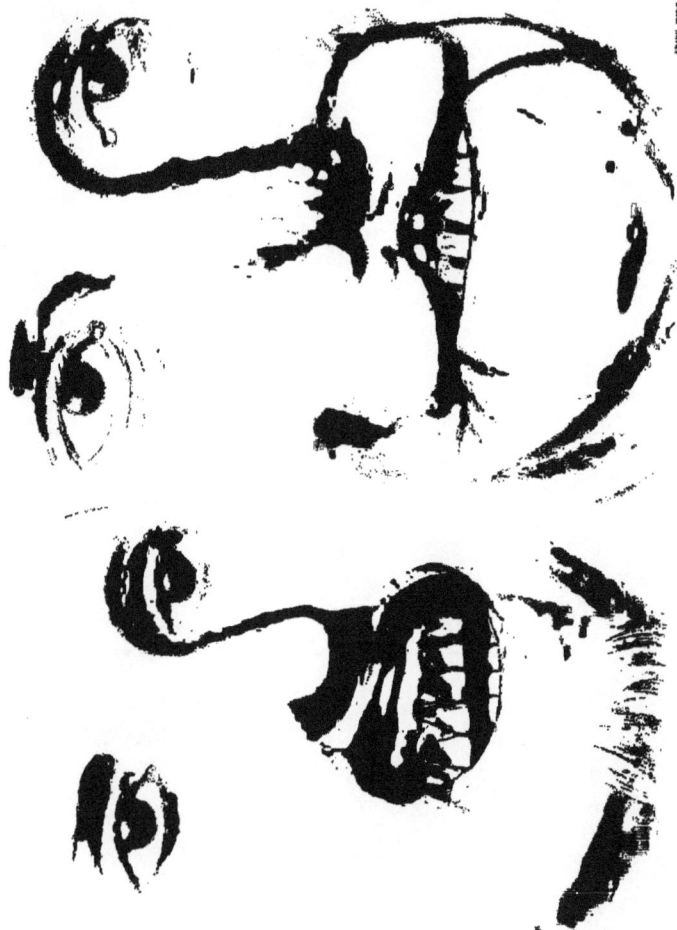

ABSENCE OF THE UPPER LIP,

THE RESULT OF APPLICATIONS FOR CURE OF ALLEGED CANCER.

PLASTIC OPERATIONS.

BY ALFRED C. POST, M.D., LL.D.,

Emeritus Professor of Surgery in, and President of the Faculty of the Medical Department of the University of the City of New York; Visiting Surgeon to the Presbyterian Hospital, N. Y. City, &c., &c.

CASE.—Chas. Gardiner, Ireland, 65, shoemaker, married, admitted to hospital April 26th, 1880. Family and previous history good.

Twenty-six years ago, a small wart appeared on upper lip, a little to right of median line. This remained twenty-two years without increasing in size and without pain. Four years ago, applications were made to it with the intention of destroying it, but they were not effectual. The wart ulcerated and increased in size, involving nearly the whole of the upper lip, and a portion of the right ala nasi ; but without giving rise to pain or constitutional disturbance. On March 31st, 1880, he went to a cancer doctress, who applied an escharotic paste which destroyed all the parts involved in the disease.

On admission, there was found to be a deficiency of nearly the whole of the upper lip, the deficiency extending from the right cheek just without the angle of the mouth to a point about twelve mm. within the left commissure of the lips. The columna nasi had entirely disappeared, and there was a deep notch at the lower part of the right ala nasi about fifteen mm. in depth. The upper gums were entirely exposed to view. The patient was unable to keep his food between his teeth, and to retain the saliva within his mouth. (See plate I, fig. 1.)

May 1st, 1880. The patient having been etherized with the usual precautions, I performed the following operation for the restoration of the lost features. (See fig. 3, 4 and 5.) I commenced by separating from the cheek the small remnant of the left extremity of the upper lip, by an incision extending through its whole breadth and thickness, leaving the flap thus separated, attached at the junction of the nose and cheek. The loose end of this flap and its anterior surface were excised, so as to leave raw surfaces for adhesion in the new position in which they were to be placed. A portion of this integument at the end of the nasal pyramid, and at the lower border of the septum

[9]

nasi were also excised, to produce raw surfaces corresponding with those of the labial flap. This flap was then drawn upward and to the right, and attached by sutures to the denuded surface of the nose, so as to form a new columna, the mucous membrane looking downward, and the denuded cutaneous surface looking upward. The position of the new columna was necessarily oblique, it being my intention to remove the obliquity by a subsequent operation, if it should be found necessary.

The jagged edge of the parts at the junction of the right cheek with the lip was then excised so as to leave an even margin, and a flap was made from the left cheek, included between two incisions, of which the lower extended outward and a little downward from the angle of the mouth, and the upper, outward and a little upward from the junction of the ala nasi with the cheek, the flap thus formed being a little larger in all its dimensions than the left half of the space to be occupied by the new upper lip. From the outer extremity of this flap, two incisions were made in the arcs of circles whose concavities looked toward the mouth, including between them a curved pedicle for the flap, about half the width of the flap itself. A corresponding flap with a curved pedicle was cut from the right cheek, the flap being made sufficiently wide, not only to supply the right half of the upper lip, but to afford a patch to fill up the notch at the lower extremity of the right ala nasi. A horizontal incision was made twenty-two

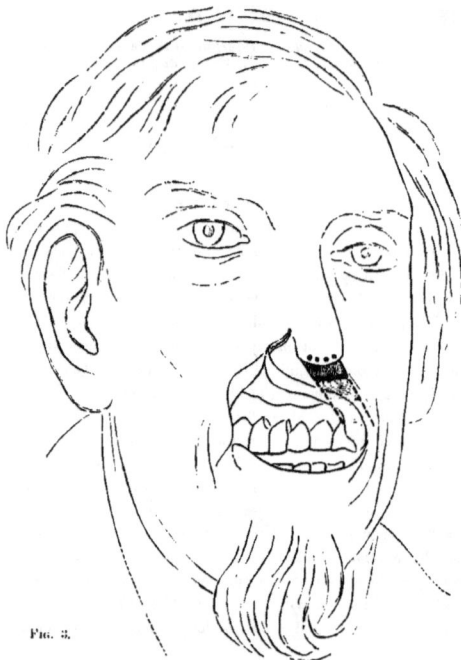

FIG. 3.

mm. in length, separating the portion of the flap designed for the reparation of the ala nasi from that which was intended for the completion of the upper lip. A thin strip was removed from the margin of the notch at the lower edge of the ala nasi, for the reception of the flap. The two flaps designed for the formation of the upper lip were then brought together in the median line, and joined by three pin sutures, and the remaining edges of the wound were united by numerous fine silken sutures. Before bringing the flaps together, as there was found to be some tension I cut across the mucous membranes of the pedicle of the right flap near its lower extremity, thinking that the mucous membrane of the flap itself would receive a sufficient vascular supply from the vessels of the submucous cellular tissue. A new vermillion border was formed for each division of the lip, by drawing the mucous membrane forward, and attaching it by sutures to the external integument.

The patient bore the operation well, being kept in a state of anæsthesia by a very moderate use of ether.

Fig. 3 exhibits the flaps which were designed to restore the columna nasi, the upper lip, and the defective portion of the right ala nasi.

Fig. 4 shows the new columna attached by sutures to the apex of the nose.

Fig. 5 exhibits the remaining flaps secured in place by sutures and the space behind the right flap, left to granulate.

May 2nd. Patient is in a comfortable condition. He has taken milk and beef-tea during the night. The wound looks well, except that the vermillion border of the right flap has a somewhat dusky appearance.

5th. One of the pins, and about half of the sutures were removed to-day. A large part of the wound has united by first intention, but the vermillion border of the right flap is evidently gangrenous.

May 10th. The remaining sutures were removed.

16th. The sloughy margin of the right flap came away to-day.

25th. Union is complete, except where the slough separated from the margin of the lip and at the posterior superior margin of the pedicle, where cicatrization is rapidly advancing.

June 22nd. Every part of the sore is healed, except a minute portion about one cm. in length, and five mm. in breadth, at the posterior and superior margin of the pedicle, where the surface is covered with a thin scab on a level with the surrounding parts. The line of union of the right flap with the adjacent parts is at its proper level, except for a space of three cm. in length, extending outward from the angle of the mouth, when the line of union is depressed, and the adjacent surface puckered.

The line of union of the left flap is level on its superior margin, and on the posterior side of its pedicle, but in other parts it is depressed, and the adjacent surface somewhat puckered. The vermillion border of the left flap is perfect throughout, and extends a little to the right of the median line. The free border of the right flap has no proper mucous covering, but the skin is inverted so as to form a fair substitute for a vermillion border, from the angle of the mouth about half way to the median line. Between the labial borders of the right and left flaps, there is a notch about two cm. long at the base, and extending upward to the height of one cm. above which the two sides of the lip are firmly but somewhat irregularly united.

The flap which fills up the notch in the right ala of the nose is firmly united, but is abnormally thick. The new columna is perfectly united with the septum and the apex of the nose, and its obliquity is much less than might have been anticipated. (See plate 1, fig. 2, see fig. 9, exhibiting the result of the first operation.)

June 23rd. I performed the following supplementary operation. I first detached the left side of the posterior extremity of the newly formed columna to the extent of about three mm. I then excised a segment from the right edge of the columna, about three mm. in breadth in the middle, and tapering toward each extremity, and brought together by fine sutures the sides of the chasm thus produced. I separated from the cheek the newly formed patch of the right ala nasi, by a deep incision, and dissected out a considerable portion of the subcutaneous adipose tissue, so as to diminish the thickness of the flap, and to reduce it to its proper level. The patch was secured in its place by fine sutures. I then dissected out the thin cicatricial tissue on the right side of the middle of the lip, so as to leave the two divisions of the lip each with a straight parallel margin. The incisions were made in such a manner that the left or longer flap was bevelled at the expense of its mucous surface, and the right or smaller flap was bevelled at the expense of its cutaneous surface, so that when the flaps were afterwards brought together, the left flap overlapped the right one, while the cutaneous surfaces of the two flaps were on the same plane. This was done in accordance with a plan suggested by Dr. Packard of Philadelphia in a paper read before the New York Academy of Medicine. The two flaps were both detached from the bridles which bound them to the upper jaw and to the buccal mucous membrane. I then made an incision on each side on a line corresponding with the upper margin of the lip, extending into the cheek to the distance of about four cm., and the same line

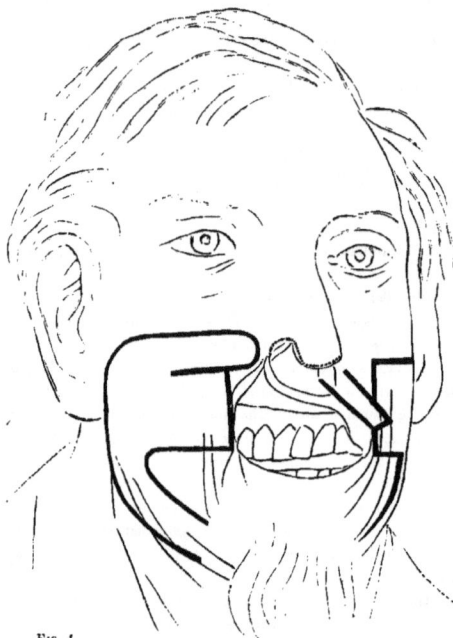

Fig. 4.

of incision was then curved downward behind the angle of the mouth, and forward along the lower lip to the extent of three cm., the flaps of the lower lip being about twelve mm. in breadth. The two flaps of the upper lip were then brought together near the median line, and were secured by two pins and a number of fine sutures. A small triangular surface on the right cheek, about twenty-five mm. in length was left to granulate. This space was lightly filled with picked lint moistened with a solution or carbolic acid, one part to forty, and the surface was

[12]

covered with lint moistened with collodion. The patient bore the operation well, and was in a good condition at its close.

July 3d. There is a considerable separation of the flaps in the median line. There is sloughing of a considerable portion of right side of lip. Surface thoroughly washed with carbolic lotion.

20th. Since the separation of the sloughs, the flaps have been supported by strips of adhesive plaster, and the intermediate space has been filled by granulation.

25th. There is firm union between the flaps by a narrow band with a notch below, and a small hole above. There is also a slight notch at the lower margin of the right ala nasi.

Sept. 9th. I performed another operation as follows. I made an incision along the base of the lower jaw, commencing at the median line, and extending to the right to a point about three cm. beyond a perpendicular line falling from the right angle of the mouth. The incision was then curved upward and backward to the horizontal level of the angle of the mouth, and thence upward and forward to the junction of the upper lip with the ala nasi, and thence forward along the upper margin of the lip into the vacant space which was left by the sloughing which followed the previous operation. This incision was extended deeply into the subcutaneous tissue along the base of the jaw, and when it passed through the cheek, it was extended through the mucous membrane into the buccal cavity. At the inner extremity of the large flaps included within the limits which have been described, a portion of the free margin of the right side of the lip was detached by an oblique incision from this isthmus, leaving a flap about twelve mm. in length, and five mm. in breadth at its right extremity and tapering to nothing at the left. A corresponding surface was prepared for the reception of this oblique flap, by dividing the left portion of the lip to a similar extent. The cicatricial tissue between the right and left portions of the lip was then dissected out. The right division of the lip, with the large curved flap to which it was attached, was then drawn to the median line, and attached by two pin sutures and a number of fine silk sutures, in such a position that the small oblique flap connected with the right segment of the lip overlapped the denuded portion of margin of the left segment, and was placed in accurate coaptation with it. In this way, the whole of the face margin of the lip was made continuous, without any appearance of a notch. After securing the lip in its place, the whole circumference of the flap was closely attached by sutures to the surrounding parts, without undue tension at any point. All the parts were then washed with carbolic acid lotion, one part to forty.

15th. Pins and sutures removed. Union throughout, except at two points, one a cm. from the columna, and a second smaller surface a little further to the right. The parts were supported by strips of adhesive plaster. The result of this operation was entirely successful, but when the parts were healed, there was a marked deficiency of the vermillion border of the right side of the upper lip, and a superfluity of the corresponding border of the lower lip.

On the 14th of October I operated, with the hope of remedying this condition.

A point was selected about seventeen mm. above the right angle of the mouth, and another point at the junction of the skin with the vermillion border of the lower lip, two cm. to the left of the right angle of the mouth. At each of these points a small pin was inserted through the skin, while a third point was selected in the cheek six mm. below a horizontal line extending back from the second point indicated and four cm. to the right of the angle of the mouth, and a third pin was inserted at that point. These pins were designed to mark the outlines of a triangular flap, whose base embraced the angle of the mouth, and whose apex corresponded with the point indicated by the third pin inserted into the cheek. Introducing the index finger of my left hand into the buccal cavity as a guide, I inserted the point of a Beer's cataract knife, at the point indicated by the first pin, into the cavity of the mouth, and made an incision to the point indicated by the third pin. I made another incision in the same manner from the third to the second pin, and thus separated the triangular flap from all its connections except around the angle of the mouth. From the point indicated by the first pin, I made another incision vertically upward to an extent corresponding with the space between the first and third pins. This incision was carried to the periosteum, and from the tension of the parts involved, the edges receded so as to make a triangular chasm, adapted to the reception of the triangular flap. The flap was then turned edgewise until its apex was received into the apex of the triangular chasm, where it was firmly secured by a pin suture. The flap thus transplanted carried with it a considerable portion of the vermillion border of the lower lip, so as to bring it into line with the vermillion border of the upper lip. The flap was then secured by fine sutures to the edges of the triangular chasm, the outer line of the triangle being nearly vertical, while the inner line passed obliquely inward and downward toward the columna nasi. The edges of the chasm, from which the flap had been cut, were brought together with one pin suture and a number

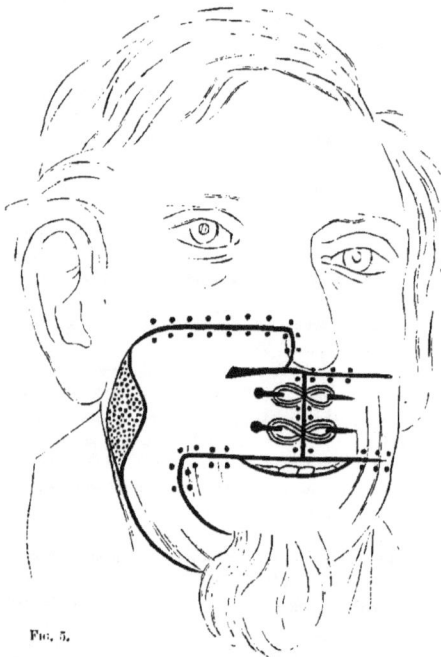

FIG. 5.

of fine silk sutures, the line of union extending from the outer side of the base of the flap in a direction downward and outward. When the wounds were all closed, it was found that the superfluity of the lower lip was entirely overcome, while there was ample material provided for the reconstruction of the upper lip. The flap which had been transplanted appeared more prominent than the surrounding parts, making the right side of the lip thicker than the left, which was the reverse of its previous condition.

The pins were removed on the 15th and 16th, and the sutures on the 23d. The wound healed perfectly, leaving the right angle of the mouth elevated above its proper level to the extent of fifteen mm. The right cheek and the corresponding angle of the mouth were adherent to the maxillary bone.

On the 11th of November I endeavored to overcome these defects by the following operations. I divided the adhesions connecting the soft parts with the bone, and then cut a triangular flap with its apex above, and its base including the angle of the mouth, carrying the incisions into the buccal cavity, and drawing down the commissure of the lips until the angle of the flaps had descended about twelve mm. below the point from which it had been detached. This flap was fixed by sutures in its new position, and the sides of the space from which it had been taken were fixed in the same way. An incision was made above the portion of the vermillion border which had been raised from the lower lip, and a raw surface was made upon the corresponding edge of the upper lip to receive the flap thus raised, and this detached portion of vermillion border was then drawn across towards the median line, and secured by sutures in its new position. The position of the angle of the mouth was improved, but it was still above its proper level. To assist in bringing it down I cut another trian gular flap below the angle of the mouth, its base including the commissure of the lips, and its apex extending toward the base of the lower jaw. Below the angle from which the apex had been cut, I made a straight incision downward to the extent of fifteen mm. and separated the sides of this incisions from each other, so as to form a new triangular flap. I then passed a loop of thread through the flap a little above its apex, and made traction so as to draw the apex of the flap down to the apex of the new triangular chasm. While this traction was made the flap was secured in its place by sutures, and the angle of the mouth was brought nearer to its proper position, but it still remained a little higher than that of the opposite side.

The sutures were removed on the 15th, and the wounds healed by first intention.

December 9th. The right angle of the mouth being considerably nearer to the median line than the left, I performed another operation for the purpose of extending the commissure outward and backward. To accomplish this object, I made an incision through the upper lip at the junction of the vermillion border with the skin, extending through the whole thickness of the lip into the buccal cavity, beginning about one cm. from the angle of the mouth, and carried around the angle along the lower lip, terminating at a distance of eighteen mm. from the commissure of the lips. I then made another incision commencing three mm. below the angle of the mouth, and extending fifteen mm. horizontally outward and backward through the whole thickness of the cheek, the anterior and internal extremity of this incision corresponding with the wound by which the vermilion border was detached from the cheek. A blunt hook was then introduced through a portion of the vermilion border detached from the lower lip, seven mm. on the inner side of what had previously constituted the angle of the mouth, and was drawn through the horizontal incision in the cheek, and the corresponding portion of the detached vermilion border was fixed in that position by a bead suture extending through the cheek eighteen mm. beyond the outer extremity of the wound. In this manner, the angle of the mouth was extended outward, and provided with vermilion border, mainly at the expense of the lower lip. The flap thus transplanted was secured in position by fine silk sutures.

23d. I made another effort to bring down the upper lip to its proper level. I made a horizontal incision immediately below the nose, extending on each side to a point vertically above the commissure

[15]

of the lips, and thence continued on each side downward and outward to the extent of six cm. beyond the angles of the mouth. These incisions were carried through into the buccal cavity, and the lip was brought down into its normal position. I then made an oblique incision through the lip, commencing at its upper part on the left side of the median line, and extending down to the vermilion border one cm. to the right of the median line, and the parts were then united by sutures, so that the relatively superfluous vermilion border of the left side overlapped the margin of the right side, a raw surface having been made for its reception. An attempt was then made to close the chasm made by the depression of the right side of the lip, by dissecting a flap from the upper part of the cheek, with a pedicle curved downward and outward to a point twelve mm. in front of the lobe of the ear. But when the flap was dissected from the subjacent parts, it was found that it could not be brought over to the internal limit of the chasm which it was designed to fill. In order to bring the flap into place, it was necessary to perform the hazardous experiment of dissecting the pedicle downward and inward to the base of the jaw, leaving the narrowest portion in front of the lobe of the ear only twelve mm. in breadth. The flap was then readily brought into place and secured by sutures, leaving a large chasm around the posterior circumference of the transplanted flap. The portion of the wound uncovered by integument was dressed with lint moistened with collodion. The surface of the flap was covered with lint smeared with salicylic ointment. 27th. The flap, which was at first pale, has assumed a brighter color. There is some œdematous swelling, with a burning sensation. The surface was washed with a carbolic acid lotion, one part to forty, and again dressed with salicylic ointment. 28th. A slight blush of erysipelas has appeared in the integument of the eyelids of the right side, and the flap has begun to assume a livid color. Ordered sulph. quiniæ, gr. viij., and tinct. ferri chloridi, min. x., to be given, and the inflamed integument to be penciled with tinct. iodini. 29th. The extremity of the flap, to the extent of five cm. has evidently lost its vitality. The erysipelas is spreading over the forehead, and has extended to the occiput. Jan. 5th, 1881. The space left by the separation of the slough is filling up with granulations. Feb. 11th. The contraction of the granulations, and the process of cicatrization, have gone on, until the chasm left by the separation of the sloughs has become nearly filled. March 4th. The wounds are substantially healed, and the right angle of the mouth has been drawn up nearly to the same position as before the last operation.

By my advice, the patient left the hospital, and went home to recruit his general health. I hope at some future time to make an effort to improve the position of the right oral commissure. The result of the first operation performed on this patient was very satisfactory, although it was far from restoring the perfect symmetry of the face. The subsequent operations contributed much less to the improvement of the patient's appearance than it was hoped that they would. The whole result has been the reconstruction of an upper lip which had been almost completely destroyed, the partial reparation of the notch in the right ala nasi, and the complete restoration of the columna nasi. The left side of the lip is nearly perfect, but the right side is drawn up above its proper level, and the mouth cannot be perfectly closed.

Note. This case was reported at the meeting of the American Medical Association held at Richmond, Va., May 4th, 1881, but has not been given to the public until the present time.

II. FIBROUS TUMOR OF THE FACE
(Case of Prof. Willard Parker.)

FIBROUS TUMORS OF THE FACE.

BY WILLARD PARKER, M.D.,

Professor of Clinical Surgery, College of Physicians and Surgeons, N. Y. ; Consulting Surgeon to the New York, Bellevue, St. Luke's, Mt. Sinai and Roosevelt Hospitals, &c., &c.

The accompanying plate II. (Fig. 6, 7, 8) represents the following :

CASE. In February, 1863, I. I., aged 60, a mechanic from Cortlandt, Westchester Co., N. Y., called at my office to consult me in relation to a tumor on the face. He was of industrious and regular habits, of a healthy family, and gave the following history.

Thirty years ago he first observed a swelling, the size of a pea, situated a little in front of the ramus of the jaw, below the zygoma. During the subsequent twenty years it grew slowly, attaining the size of a hen's egg—there was no pain or tenderness. Within the past ten years the growth has been more rapid, it having more than doubled in size during the past five years.

It now measures at its base eighteen inches ; at its greatest circumference, twenty-two inches. It is attached from the malar bone above, to a point upon the neck two inches below the thyroid cartilage, and, in front, from the angle of the mouth to two inches below the thyroid cartilage.

It projects from the face some six inches, its vertical diameter being about eight. It is freely movable, and its surface is covered with nodules varying in size. The whiskers cover its posterior surface.

He was admitted into St. Luke's Hospital, where I was consulting surgeon, and a consultation was called. The decision being to remove the mass, the patient readily assented and the operation was performed on February 27th.

An incision was made around the base of the tumor, which was readily dissected out from its attachments. The muscles of the neck and jaw were not interfered with, and no large vessels were encountered, though there was tolerably profuse hemorrhage from a number of smaller vessels. Some difficulty was experienced in bringing together the edges of the wound. The wound healed readily, largely by first intention, the patient leaving the hospital at the end of a month with complete cicatrization and but little deformity.

[17]

FIBROUS TUMORS OF THE FACE.

In consulting my record books I find the history of two other cases which will be of interest, I think, to the readers of this journal.

I insert them here briefly.

CASE.—A French gardener, aged 51, consulted me some years ago on account of a large tumor on left side of face, extending from the zygoma to a point about two inches below the angle of the jaw. He had discovered some twenty-one years previously a small movable tumor, which was without pain or tenderness. For fifteen years its growth was slow, giving no special annoyance. For some five years subsequently its growth was more rapid, and within the previous two years it had become vascular, and at times, painful. Apprehending that the character of the growth was changing, I advised its speedy removal. My advice was followed, and he came to the city for operation. No special difficulty attended the removal of the tumor, the hemorrhage, though considerable, being readily controlled. The patient made a good recovery, but the tumor returned within about a year's time, the patient dying of malignant growth in the cicatrix. The tumor weighed two and one-half pounds, and had assumed a malignant character at the time of removal.

CASE.—An unmarried woman, aged 40, of healthy family, presented herself with a tumor on right side of face, of some nineteen years' duration. Its history was almost identical with the two previously given. General health of patient good. Tumor removed by operation, recovery rapid and satisfactory. No subsequent history of return of disease.

Although the histology of Fibromata is understood by the medical profession, their cause has not as yet been definitely made out. It develops in all tissues; beginning as a very small tumor, its growth is slow, it is local in its character, and unaccompanied by pain, heat, or tenderness; the deformity when in an exposed part, being the only cause of complaint.

The diagnosis of Fibroma is generally easy, the only other growths liable to be confounded with it being lipoma, chondroma, cystic and adenoid growths. The malignant tumors differ from it in that they are generally painful, of rapid development, and occur in an unhealthy system. In fibroid growths the prognosis is good as regards a return of the disease, but it may develop in organs where surgical interference would jeopardize life, or when it so embarrasses vital action as to cause death.

The only treatment is removal of the growth, and the earlier, the safer. The reasons for early removal are:

1. When the growth is small the operation is less dangerous.

2. When located near important organs, or internally, its relations may be such as to render an operation difficult or even impossible.

Fibrous tumors, like all other abnormal developments, have a tendency to change of character, and may become malignant. Malignancy is, as a rule, never a *first* condition, but is in the beginning an outgrowth of some previous abnormal condition or growth.

To sum up briefly, cancer is *traumatic*, or has its origin in a previous lesion.

[18]

LAPAROTOMY

PERFORMED FOR THE REMOVAL OF A LARGE QUANTITY OF MENSTRUAL BLOOD FROM ONE HORN OF A BICORNATE UTERUS.

BY T. GAILLARD THOMAS, M.D.,

Surgeon to N. Y. State Woman's Hospital.

Mrs. A., a native of France, forty years of age, a widow, who had borne one child, entered my service in the Woman's Hospital, and gave the following history. For the past sixteen years she had suffered from what a large number of physicians who had examined her had uniformly pronounced to be a fibrous tumor of the uterus. At the commencement of that period she had spent eight months in the hospitals of Paris, and had since consulted many physicians, but without obtaining any relief whatsoever. The three distinguishing features of her case were these: first, since its development the tumor had neither increased nor diminished in size; second, it was at all times exquisitely sensitive to pressure, and especially so during menstruation; and, third, pain occurred in it during every menstrual act, so severe that nothing gave her relief except a free resort to opium. Her suffering during menstruation I have never seen surpassed, and she had become so demoralized by it that her object in entering the hospital was to have the growth removed at all hazards.

Upon examining her I found the pelvis filled by a tumor about as large as the head of a child a year old, which, as I have already said, was very sensitive to pressure. It was apparently solid, only slightly movable, and by conjoined manipulation appeared to be attached directly to the uterus. I saw no reason to differ from the diagnosis which had been heretofore made in the case, although I was very much puzzled by the existence of the three peculiar features to which I have already referred.

I dissuaded the patient from operation, but she was so much distressed at this that I got my colleagues, Dr. T. A. Emmet and Dr. J. B. Hunter to see her with me in consultation; she indulging the hope that they might differ with me in this regard, and declaring that so great were her sufferings that she would infinitely prefer a resort to surgical interference, however great the dangers might be, than to remain exposed to them. Drs. Emmet and Hunter agreed both in the diagnosis and in the propriety of refusing operation. The patient then left the institution, and I did not see her for five or six months, when she returned again urgently demanding operation. I kept her in my service for some time, and then, with regret, again dismissed her without having afforded her any permanent relief.

Two months after this she saw me at my office, and so fully described her sufferings, and so earnestly pleaded for relief that I again admitted her to the hospital, promising to remove the ovaries by Battey's method, in the hope of, in this way, relieving her of at least the greater portion of her troubles. For this operation, she entered my service during the year 1881.

I cut down through the abdominal walls, and reached a tumor which looked exactly like a fibroid. I put my fingers upon it, and was surprised to find an obscure and yet distinct sense of fluctuation, which had not been recognized through the abdominal walls. Instead, therefore, of going on with the intended operation, I introduced a cannla and trocar into the fluctuating tumor, and on withdrawing it there immediately escaped a pint or a pint and a half of menstrual fluid. It had all the characteristics of that fluid, and there could be no mistake with regard to its nature. I was very much puzzled by this, for the woman had been carefully examined, she had menstruated regularly, the uterus had been repeatedly measured, and was found to be two inches and a half in length. Taking hold of the tumor with two strong tenacula and drawing it up into the abdominal wound, I passed my hand down and discovered its relations, when at once it flashed across my mind that this was a uterus bicornial ; that the canal in the left horn was free, and allowed the escape of the menstrual fluid from that side, while the canal in the right horn was not open throughout its entire length, and consequently obstructed the menstrual discharge from that side.

The original condition of the parts was probably that represented in Fig. 9 or Fig. 10, one cervix being pervious, and the other impervious, if the uterus were originally bicornate as represented in Fig. 9, or else no cervix existing if the organ were originally unicornate as represented in Fig. 10.

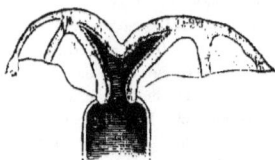

FIG. 9.— BICORN UTERUS.   FIG. 10. —UNICORN UTERUS.

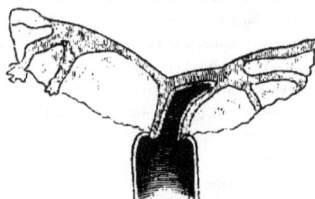

Under such circumstances, one uterus, or rather one horn, discharges menstrual blood ; in the other, that fluid, secreted by the endometrium, accummulates and creates a tumor presenting many of the features of a fibroid.

When this thought suggested itself I was able to account for all the peculiarities of the case: First, the fact that the tumor had the appearances of a fibroid, and gave such agonizing pain at each menstrual period ; second, the fact that the tumor remained at about the same size, not growing larger, as a fibroid would do, although it was not at all impossible for a tumor of this character to have become larger by gradual distension ; and, lastly, the fact that the patient had comparative immunity from pain between the menstrual periods. I now found myself in an unfortunate dilemma, for had I proceeded to remove the ovaries, blood would have escaped from the wound into the abdominal cavity, and would very likely have set up fatal peritonitis or septicæmia. Hence, I adopted a course which struck me, as under the circumstances, the only one which would meet the emergency.

The following diagram, Fig. 11, will show the condition of affairs at this stage of the operation. I caught hold of the uterus on one side, and Dr. Ward on the other, each with a strong tenaculum, and dragged it firmly up into the abdominal wound, and passed two strong knitting-needles through the tumor and laid them on the abdominal walls above and below the point of puncture. I then passed two sutures deep down, and fastened the tumor in the abdominal wound, and left a tube in for drainage and irrigation. I remarked to the spectators present, that the woman would almost surely suffer from septicæmia, and this prediction was fully verified; but by having the cavity constantly irrigated with carbolized water, and controlling the temperature by Kibbee's method of affusion, she recovered. A day-to-day history would accomplish nothing in increasing the

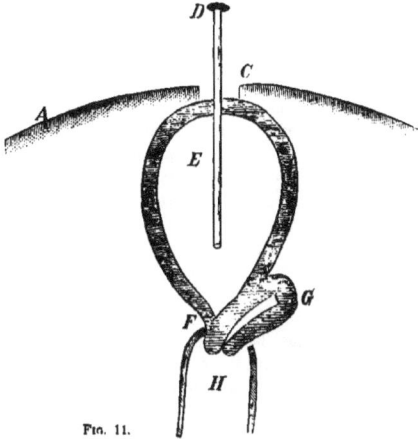

FIG. 11.

A. B. Abdominal Walls. C. Section in median line. D. Canula. E. One uterine body distended by retained menstrual blood. F. Trine uteri, impervious on one side. G. Uterus with pervious cervical canal. H. Vagina.

interest of the case, hence, I spare the reader a recapitulation of it. It suffices to say that the uterine cavity was thoroughly washed out with carbolized water every four or five hours; the temperature, which rose to 106°, kept in the neighborhood of 100° by affusion; and quinine and opium used freely; and the patient treated in all respects as she would have been after ovariotomy.

At the next period, menstrual blood escaped simultaneously from the vagina and the abdominal wound, the drainage tube not having been removed from the latter. Never have I known more complete relief ensue from any operation than from this one. The patient constantly expressed herself as entirely relieved, and is so well satisfied with her present condition that she is entirely unwilling to consider a procedure to which I shall soon allude for improving it.

For her, however, to continue in her present state of comfort, it is evidently essential that the abdominal opening shall be kept free until the menopause. To accomplish this, as soon as the menstrual period was over, I had constructed a solid glass rod represented in actual size by Fig. 12.

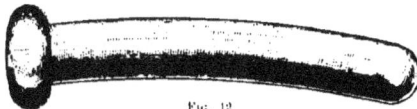

FIG. 12.

21

This, the patient wears constantly, except at menstrual periods, keeping it in position by a girdle which presses upon its head, and at the same time sustains the parts about the incision.

She has now left the hospital, but reports to me occasionally, and has been instructed how to allow the free escape of menstrual blood by the geno-pectoral position, and how to wash out the cavity with carbolized water, in case of any septic symptoms attending upon or following menstruation. This last manœuvre she has frequently practiced, and perfectly understands. Fig. 13, although a rough diagram, represents very accurately, I think, the present state of affairs.

When this operation was adopted, I felt that it had helped my patient and myself out of a very difficult dilemma, but at the present time I do not feel at all satisfied with the *status rerum.* Should the patient not become pregnant, it is highly probable that all will go well with her until the menopause, but should pregnancy occur in the left horn, the right will assuredly be torn away from its abdominal moorings, and a fatal issue would occur.

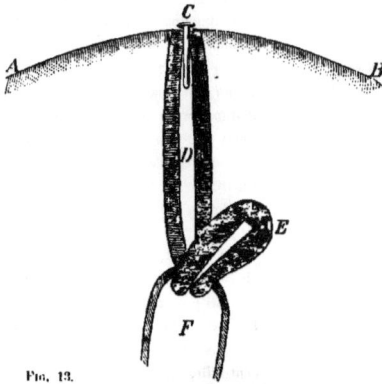

Fig. 13.

A. B. Abdominal wall. C. Glass plug. D. Uterine body fastened to the abdominal wall. E. Uterus with pervious cervical canal. F. Vagina.

At the moment of operation, and since that time, the propriety of penetrating the impervious cervical canal of the right horn, keeping it permanently open, closing the upper opening, dropping this horn into the pelvic cavity, and then closing the abdominal wound, has been carefully considered.

As is so often the fact, under similar circumstances, the advisability of this course will very likely be immediately determined upon by many who have not had an opportunity of observing the case. To those who have watched it with keen anxiety, through its various phases, much more difficulty will attend the decision. The patient is past forty; should conception occur the propriety of checking utero-gestation would be quite evident; the patient, who has gone through with a great deal of suffering, strenuously objects to interference with a condition which is perfectly satisfactory to her and the dangers attendant upon the steps referred to would be very considerable.

So many years have elapsed since pregnancy occurred with her, that I think it highly improbable that it will now take place after the fortieth year. Should it do so, I should feel myself called upon under present circumstances to put a stop to its progress. This, however, I should feel justified in doing only once; having once resorted to it as a therapeutic resource, I should feel it my duty to urge upon her a resort to those further surgical steps which I have mentioned.

[22]

SEPARATION OF THE LOWER EPIPHYSIS OF THE FEMUR.

BY JAMES L. LITTLE, M.D.,

Professor of the Principles and Practice of Surgery, University of Vermont; Professor of Clinical Surgery, University of New York; Surgeon to St. Luke's Hospital, New York City, &c., &c.

Plate III. illustrates this somewhat rare form of injury. The separation is complete except at the inner and external surface, where a small piece of the shaft was chipped off and remained attached to the epiphysis. This small piece was lost before the sketch was made by the artist, Mr. George C. Wright.

The history of the case is as follows: On April 18th, 1865, a boy, aged eleven, while hanging on the back of a wagon, stealing a ride, had his right leg caught between the spokes of the wheel while in motion. The result of this sudden and forcible twisting of the leg was to cause a diastasis of the lower epiphysis, and the forcing of the lower extremity of the shaft through the soft parts on the upper and outer part of the popliteal space, the upper fragment completely overlapping the lower. The protruding part of the shaft was cleanly stripped of its periosteum for a space of about three inches. The hemorrhage was quite free, but no ligatures were necessary. The fragments were readjusted by strong extension of the limb, and flexion of the leg upon the thigh while patient was under the influence of ether. A dull cartilaginous crepitus could be felt when the fragments were moved upon each other. The wound did not appear to communicate with the knee joint. After reduction, the limb was placed on a double inclined plane. The following day, the upper fragment was found to be displaced. The patient was again placed under an anæsthetic, and it was then found that the fragments could be kept in place only by extreme flexion of the leg. This position was maintained by binding the leg to the thigh by means of strips of adhesive plaster. This apparently constrained position was not uncomfortable to the patient. On the thirteenth day alarming secondary hemorrhage took place while dressing the wound. This was controlled by a tourniquet, and as soon as the patient rallied from the effects of the loss of blood, amputation of the thigh at its lower third was performed. The boy made a good recovery. The hemorrhage was found to have come from an injury to the anterior tibial artery near its origin. The plate shows the situation of the wound.

Dr. Hamilton, in his work on Fractures and Dislocations, reports three additional cases of this form of injury. In one case, an abscess formed and amputation became necessary; in another, the boy recovered with a straight leg, but with complete anchylosis of the knee joint and short-ening of the limb three-quarters of an inch.

The lower epiphysis of the femur does not become united to the shaft until the twentieth year. So that at any period under twenty it is possible that this form of injury may occur. In several cases that have been reported, the accident happened in the same way, the leg being caught in the spokes of a wheel while in motion. This, like the separation of epiphyses of other bones, rarely, if ever takes place without a chipping off of a small portion of the shaft. In some cases there is a partial separation of the epiphysis with a fracture of the adjacent portion of the shaft, leaving a large portion of the latter attached to the condyle. In the museum of St. George's Hospital, London, there are two specimens of this last variety, in one of which the lower fragment is in two pieces.*

In cases where the injury is not compound, displacement of the upper fragment, downwards and backwards, may be so great that the nerves and vessels may be stretched over its edges. In one case, under Dr. McBurney's care, at St. Luke's Hospital, the internal popliteal nerve was so stretched, the pain so intense, and the deformity so great that amputation was resorted to. The case was of two or three months' standing, and strong bony union had taken place. The patient died of tetanus.

In this case, and in my own, as well as in the case reported by Dr. Hamilton, the periosteum was completely stripped from the end of the shaft. Jonathan Hutchinson says: "This is, I believe, invariable in cases of detachment of an epiphysis with displacement." The periosteum is always left as a sleeve in connection with the epiphysis and the shaft is denuded.† In Dr. McBurney's case, although it was not compound, and union had taken place before amputation, yet the lower end of the shaft was found to be bare of periosteum. The vitality of the bone was not interfered with by this loss.

In this case, as well as in two cases described by Hutchinson, the epiphysis was drawn into the flexed position by the action of the gastrocnemius muscle. This condition seems to be difficult of rectification, and coaptation of the fragments can only be maintained by placing the limb in the position of extreme flexion.

The results of this injury when compound are unfavorable. Secondary amputation has been frequently resorted to in consequence of the profuse suppuration and difficulty of keeping the fragment in position.

* Surgical Treatment of the Diseases of Infancy and Childhood. By T. Holmes, 1869. Page 259.
† Illustrations of Clinical Surgery. Vol. II. Page I.

DISLOCATION OF THE COLUMNAR CARTILAGE OF THE NOSE.

BY F. H. BOSWORTH, M.D.,

Lecturer on Diseases of the Throat in the Bellevue Hospital Medical College.

Lying immediately below the cartilage of the nasal septum, and parallel with its lower border, is a small oblong plate of cartilage not usually mentioned in the text-books of anatomy. It lies immediately beneath the integument of the columna, and can be easily grasped between the thumb and forefinger. It seems designed to give firmness and shape to the columna, and also probably aids in supporting the tip of the nose.

Within the past year, two cases of lateral displacement of this cartilage have come under my care, which seem to possess an interest from being somewhat unique and also from the simplicity and success with which the deformity was rectified.

The first case was that of a gentleman from Illinois who consulted me in March last, reporting that two years before he had noticed that something was growing in his right nostril. He had paid little attention to the matter at the time, but the condition had progressively grown worse until, at the time he came to the city, the deformity had increased to such an extent that there was not only a considerable degree of closure of the nostril, but the nose presented such a noticeable loss of symmetry as to become a source of no little mental distress. In fact, the matter annoyed him so much that he came to New York with the sole purpose of seeking relief by some surgical operation. When I saw him the nose presented much the appearance illustrated in the accompanying sketch. The general facial expression, however, was even more noticeable than is suggested by the cut; it was very peculiar, and indeed almost sinister, and I was not surprised at his anxiety for relief, especially as he said the deformity seemed to be increasing. The case was unique to me, and I was at first somewhat puzzled to determine its nature. On examination, however, I found that the columnar cartilage was displaced laterally, and at the same time tilted upward in such a manner that its posterior angle protruded into the nostril in an upward direction. It was easily grasped between the finger and thumb, and its outlines sufficiently showed that there was no deformity of the cartilage, or new growth. Furthermore, it could be restored to its proper position by pressure, but immediately resumed its abnormal position when the pressure was relaxed.

The second case was of a gentleman of this city, and differed in no especial manner from the first, except that the dislocation was to the left, which would suggest a probable explanation of the affection. In the first case the gentleman was left-handed, while in the latter he was right-handed. Probably the deformity was caused, primarily, by pressure of the thumb in using the handkerchief. Subsequently, as the nostril became partially closed, more vigorous efforts at clearing the occluded passage would be attempted by closing the opposite side by the thumb, and thereby the deformity be aggravated.

The operation in each was the same, and consisted in the removal of the cartilage. This was done by making an incision along the edge of the protruding mass with a gum lancet, which I found to

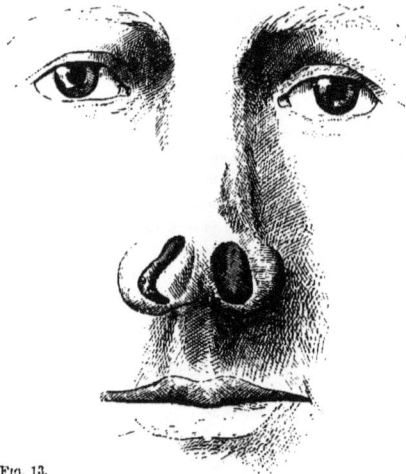

Fig. 13.

answer an admirable purpose, and dissecting down until it was sufficiently cleared to enable me to seize it with a pair of rat-tooth forceps. The dissection was then easily completed, and the cartilage extracted. A small elliptical piece of redundant muco-cutaneous membrane was then cut out with a pair of scissors, and the wound closed with two fine sutures. In each case the parts healed by first intention, the sutures being removed on the second day. No anaesthetic was used, and the operation was not especially painful. The results were eminently satisfactory, the deformity being completely removed, and absolute symmetry restored.

[26]

IV. FACIAL PARALYSIS

CASES OF DR. SAMUEL SEXTON

FACIAL PARALYSIS

OCCURRING IN CONNECTION WITH AURAL DISEASE,

WITH TWO ILLUSTRATIVE CASES.

BY SAMUEL SEXTON, M.D.,

Aural Surgeon to the New York City Eye and Ear Infirmary.

CASE I.—Facial paralysis of the right side-caused by necrosis of the petrous bone, occurring in a case of chronic purulent inflammation of the middle-ear:—see Plate IV., two upper figures, the patient in repose, and the same patient endeavoring to laugh.

The case was a woman, nineteen years of age. When about ten years old she experienced buzzing in the right ear. Coming for treatment in November, 1880, she stated that the ear never discharged until six months previously; commencing after "pains and a gathering" in the ear. Four months ago she caught a severe cold in the head, and one week afterwards she found on getting up in the morning that the right side of the face was paralyzed; the mouth was drawn to the left, and she was unable to close the right eye. She was not suffering pain when she came for treatment, but previously she had suffered much from neuralgia in the right temporal region, perhaps partly from dental caries. For the past four months the attacks have been of a more distinctly paroxysmal character. For the past two months she was never free of vertigo, and was in constant fear of falling backwards; her gait is staggering. When pressed for a statement respecting the duration of the aural disease, she admitted that the ear had always had a bad odor.

On examination, a large polypus was seen to almost fill the right external auditory canal, and the probe detected the presence of a detached sequestrum of bone deep in the canal. The polypus, which was attached to the superior posterior wall of the canal, near its outer extremity, was removed by the snare, and the sequestrum was immediately afterwards brought away by the foreign body forceps. The latter came away with difficulty, although the external auditory canal was fortunately very large. The sequestrum, when examined, proved to consist of a plate of irregularly rounded bone about one-fourth of an inch in diameter, one portion of which was very thin, and the other nearly a quarter of an inch in thickness. This sequestrum was examined by Prof. J. D. Bryant and myself, and is believed to consist of the roof of the tympanum and a portion of the mastoid body just external to the hiatus

Fallopii. A ridge on the specimen corresponds to the anterior ridge of the groove on the superior border of the bone located beneath the superior petrosal sinus. The reticulated arrangement in the cavity of the specimen has the general arrangement of the reticulæ of the antrum mastoideum ; besides which, the relations of the compact and cancellous tissues are such as to exclude its having been located elsewhere. The superior surface of the sequestrum, where it came in contact with the dura mater, was roughened by caries.

After the removal of the sequestrum, another large polypoid mass was taken away from the ear, leaving a very large cavity at the junction of the canal and the posterior wall of the tympanum. At the bottom of this cavity was a mass of granulation tissue, which, under the use of powdered acidum boracicum, soon disappeared, and the discharge ceased. The facial nerve, when tested by Dr. McBride, did not respond to either galvanism or faradism, although at first the muscles of the face gave degenerative reaction.

In this case, the facial nerve was undoubtedly impaired before the chorda tympani was given off, for there was very decided modification of the sense of taste on the right side of the tongue. The palate was unaffected. The palsy has been treated by the galvanic current, but with slight improvement. Hearing was not entirely lost. The patient is at present free of vertigo and tinnitus aurium, but has occasional headaches, owing to a rather broken-down condition, and the continuance of oral and naso-pharyngeal irritation.

Case II. Bilateral facial paralysis occurring in a case of sudden deafness from syphilis :—see Plate IV., two lower figures : the patient in repose, and the same patient endeavoring to close the mouth and eyes.

This patient was a man 40 years of age, who contracted syphilis in the summer of 1879, and was treated in Charity Hospital. Three months later, having taken a severe cold, he had paralysis of the *right* side of the face. Three months after this attack—and six months subsequent to the syphilitic infection, he was again very much exposed in a cold rain-storm, and went to bed with a severe cold ; on getting up the following morning he experienced so much vertigo that he could only walk with difficulty. While eating his breakfast, he found that he could open his mouth only wide enough to admit a spoon between his teeth. A facial paralysis of the *left* side had now occurred, and the inability to get food into the mouth was owing, doubtlessly, to the unique employment of the unaffected muscles used in mastication.

The patient at this time experienced severe pains in the right side of the head. Soon after the experience above related, as taking place at breakfast, while trying to converse with a friend, he found himself to be perfectly deaf in both ears, since which time he has been unable to hear a single word, however loudly spoken. The patient was not conscious of any deafness following the first attack of paralysis on the right side.

When the patient first experienced difficulty in opening the mouth, he fancied that he had "lock-jaw," and he then ascertained that instead of the face being drawn to the left, both sides were now alike. Following both these paralytic invasions he experienced distressing tinnitus aurium, which continued up to the time he was seen—some eighteen months after the initial attack. He also suffered greatly from pains in the head and vertigo until a short time before I saw him. He could not, of course, either whisper or whistle. No treatment was attempted.

FACIAL PARALYSIS.

The view of this patient when the face was in repose gives the characteristic facial expression in this affection ; the eyes have a horridly staring look, while the entire face is an expressionless blank. When trying to explain the symptoms of his case the difficulty experienced in enunciation, together with the nasal tone and collapsing of the nostrils—the latter preventing the entrance of air into the nose—caused the patient to exert himself in a most painful manner, yet the face gave no evidence of the struggle taking place. The absence of nearly all of the teeth rendered articulation still more difficult. When an effort was made to close the mouth and eyes, the former was accomplished by the action of the temporal, masseter and internal pterygoid muscles; the patient was, however, inclined to use his hand when requested to bring the jaws together. It was not possible to close the eyes, but he was able to roll them upwards and inwards, the lower lid remaining inactive, the upper lid dropping down slightly by its own weight ; the effect of the display of the lower portion of the cornea between the widely separated lids was ghastly in the extreme. Fortunately for these cases, the levator palpebrae not being supplied by the facial nerve, the upper lid can be raised from the eye.

The cause of the paralysis in this case is somewhat in doubt. There is a possibility that the morbid process which gave rise to it may have been at the base of the brain ; if so, it was probably syphilitic. There are reasons, however, for believing that the cause was peripheral, the most important of which are the aural symptoms ; the attacks followed colds during which there were pains in the neighborhood of the ear and disturbances of the functions of both the transmitting and perceptive regions of the ear. The exact seat of the lesion, if peripheral, cannot be told, for there were but slight morbid changes in the ear perceptible to the eye, and the patient did not return again to have the sense of taste, etc., tested.*

The symptoms accompanying Bell's paralysis are so familiar to the profession that I shall only venture to offer the above cases as a contribution to the literature of the subject, and as pointing to diseased processes about the ear as a frequent cause, leaving the discussion of the strictly neurological aspect of the disease to those who make a special study of the subject. The nearness of the aqueductus Fallopii to the middle-ear, which is well known to be specially obnoxious to colds, would seem to account for the frequency of paresis of the facial nerve in disease of the ear rather than cold affecting the nerve after it has left the stylo-mastoid foramen. It is well known to otologists that grave aural disease may exist for a long time without implicating the facial nerve : —and is it not probable that the Fallopian canal may, in some instances, from its defectiveness, afford inadequate protection to the nerve?

* For other features of this, especially as regards the hearing, see American Journal of Otology, vol. II., p. 304.

[NOTE. In the artotype illustrations the *right* and *left* sides are transposed as in a mirror.]

A RARE FORM OF CORNEAL OPACITY.

BY THOMAS R. POOLEY, M.D.,

Assistant Surgeon, New York Ophthalmic and Aural Institute.

This disease has been described under various names, as Symmetrical Opacity of the Cornea, Transverse Opacity, Transverse Calcareous Film of the Cornea (Nettleship), and Ribbon or Band-Shaped Keratitis.

It is of rare occurrence, and the best description of the disease, with an analysis of twenty-two cases, may be found in the Archives of Opthalmology, Vol. viii., No. 3, p. 293, *et seq.*, contributed by Edward Nettleship, of London.

The following case, of which the drawing is a beautiful illustration, came under my notice several years ago, and is one of the few cases of the kind which I have seen.

The patient was a German, about fifty years of age, and of apparent good health. I remember, especially, that he had no tendency to gout or rheumatism. I did not question him as to syphilis. In the right eye, which is represented in the cut, the disease began a number of years ago, and was first noticed as an opacity at the margin of the cornea, on each side, which gradually grew towards the centre, and then began to very seriously affect his sight.

On looking at the eye without raising the lid, a transverse grayish-brown opacity seemed to occupy the entire palpebral fissure. The band of opacity does not exactly follow the horizontal meridian, but crosses the cornea rather obliquely. It is much broader at the margins of the cornea, and gradually grows narrower toward the centre. By oblique illumination the epithelium of the cornea does not show any alteration, but appears quite normal. The opacity has a stippled appearance, and in places shows fissures or cracks. The opacity does not quite reach the borders of the cornea, although it is incorrectly represented to do so in the cut.

The pupil is almost concealed, except just the outlines of its upper and lower borders, which become a great deal more apparent when a solution of atropine is put in the eye, and it is then seen, as shown in the figure, that there is a synechia at its lower margin. Objects held above or below could be seen, but I have no accurate record of the acuteness of vision. In the left eye there were two patches of opacity, just beginning near the corneal margin of both sides. This eye was in other respects normal, and vision good.

I proposed to the patient an iridectomy on the left eye, which would certainly have very much improved his vision, but he declined all treatment, and I never saw anything further of the case.

The above case is quite a typical one, and illustrates very well the usual course of the disease. It seems quite evident to me that the disease should not be considered as an inflammatory one, and hence I would not speak of it as a keratitis. On the whole, I prefer the name suggested by Nettleship. The opacity may begin either as a single patch of opacity in the centre or at the sides of the cornea, or as in my case, as a patch at both sides, which gradually coalesce. The form in which it begins in the centre and grows towards the margin is rarer. The opacity is beneath the epithelium, which remains smooth, shows the usual bright reflex, and ulceration is never present. The course of the opaque band is not exactly transverse, but always slopes a little from within outwards and downwards, and the uncovered parts of the cornea are sharply defined. Nettleship says that the majority of cases remain free from complication for many years. But in all of the cases which I can remember seeing, three in number, complications of some sort existed. In the one now reported, the posterior synechia shows that iritis was present at some time. In one other case, which I saw some years since in the New York Ophthalmic and Aural Institute, there was chronic glaucoma in both eyes, with great increase of tension of the eye-ball. In the other, a girl of twenty, whom I saw in 1878 at the Columbus Blind Asylum—by the way, the youngest subject in which I have seen the disease, and younger than any of the cases analysed by Nettleship, the youngest being twenty-five—the second eye was lost by cyclitis. A case has, however, recently been reported by Lewkowitsch, in the Klinische Monatsblätter für Augenheilkunde, vol. xix., p. 250, of ribbon-shaped keratitis in an eye with a large and immovable pupil without any other abnormality. The other eye was normal. The writer speaks of the case as unusual because of the absence of other abnormalities.

Although the disease usually begins first on one cornea, it always becomes symmetrical in the end. Facts as to this statement are given by Nettleship in his cases. As to the character of the opacity—scrapings removed from the cornea and submitted to microscopical examinations have been found to consist of phosphate and carbonate of lime. Both Mr. Dixon* and Mr. Bowman† succeeded in greatly improving sight by the removal of the opacity in this way from the centre of the cornea. The pain felt by the patient during the operation proved the fact that the epithelium was not destroyed by the disease. This plan of treatment, the solution of the deposit by some chemical substance, and the operation of iridectomy, embrace the various methods of treatment which have been suggested. I confess my preference for the last-named plan, and would especially use it as the best when either synechiae from iritis or glaucoma exists as a complication, because of the well-known benefit which the operation is known to have in such cases.

Nettleship gives a careful analysis of the cases upon which his paper is based with reference to their aetiology. But it seems to the writer, even from the analysis which he gives of the facts as to the constitutional state and morbid tendencies, his view that the affection is closely allied to gout is hardly carried out. Of the twenty-two cases most proved to be in good health, when seen, four were said to

* See Dixon's Diseases of the Eye.
† Bowman on the Parts concerned in the Operation on the Eye.

be thin, dry-fibred and sallow, one only plethoric. In two there was a family history of consumption. There was a definite history of gout either in the patient or of his father in four cases, and in one other the patient though not gouty was a painter. In five other cases, although no gout was present, the patients had other changes in the eyes, which are, the author says, closely allied to gout, iritis, glaucoma, hemorrhagic retinitis. One died with granular disease of the kidneys, hypertrophy of the heart, and pulmonary apoplexy, with widely diffused and abundant atheroma of the arteries, at the age of fifty-three. He had never had gout, and the absence of gouty changes is especially mentioned in the notes of the post-mortem. In five more there is no mention made of the state of the patient's health.

Nettleship follows this analysis with some facts, and queries what he thinks may be considered *in favor of local or constitutional causes* to these we can only just allude. In favor of local cause (*a*) The opacity never invades those parts of the cornea which are habitually covered by the lids. (*b*) A margin of cornea at each end always remains free from the opacity. (*c*) It will be worth while to inquire whether, from decrease in sensibility of the cornea, such patients sleep with the eyes partially open or wink less often than other persons. Such peculiarities would allow of a freer evaporation from the exposed part of the cornea, and might therefore lead to collection of any solid residue at this place. (*d*) Is it due to a superficial *inflammatory* change leading to calcification? (*e*) Can the disease be partly explained by natural differences in the closeness or permeability of the corneal tissue, or in the thickness of its epithelium in different persons?

In favor of a constitutional cause. (*a*) Although the opacity forms in an exposed part of the cornea, it occurs in so few persons that there must be some special conditions added ; such as alteration of the eye fluids and therefore probably of the blood.

(*c*) The history of gout in one case, and in the parents of several more, and the occurrence of diseases usually attributed to gout, seem to point to it, or to the excess of uric acid in the blood as a cause.

(*d*) Is any corresponding (not necessarily identical) change met with elsewhere in the skin?

(*e*) Is the disease met with in the lower animals?

(*f*) In what relation does the corneal change stand to glaucoma and iritis when these occur?

In concluding his article this author distinguishes between the present disease and the formation of a stripe of opacity of a somewhat similar character which occurs in eyes which have been for a long time blind. The pathological appearances, too, are different in the secondary band shaped opacity. Goldzieher found them to consist of colloid formations in the superficial layers of the cornea, irregular thickening and degeneration of the epithelium and the presence of masses of fat in the deeper layers (Hirschberg's Central Batt., Jan., 1879. Quoted by Nettleship).

Reference has been made to but a few of the points of interest in this peculiar affection, and those which I have thought most likely to interest the general reader. To those who may desire to become more familiar with the subject, I would recommend a perusal of Nettleship's paper, which is a most exhaustive study of the subject. His paper has an appendix with the names of authors and an abstract of all recorded cases in the order of their publication. I have only been able to add four cases to his list, the one here reported, two others observed by myself, mentioned in the paper, and that of Lewkowitsch.

OVARIAN PREGNANCY.

BY ISAAC E. TAYLOR, M.D.,

Emeritus Professor of Obstetrics and Diseases of Women and Children, and President of the Faculty of the Bellevue Hospital Medical College, New York.

This pathological specimen of aberrant pregnancy came under my notice some years since.

A female between twenty-four and twenty-five years of age, married only a few months—general health good—regular every month, had passed her last period about two weeks, and considered herself four or five weeks pregnant. The ordinary signs of gestation, though slight, existed. At this time she was affected with severe vomiting, accompanied with intense pain over the uterine and left ovarian region. Under the treatment adopted she was materially relieved. The following day the uterine and ovarian pains returned, accompanied with fainting, when sudden collapse ensued, and the patient died.

Autopsy.—The uterus was larger than in the unimpregnated state—was two and three-quarters inches in length, and one and three-quarters in breadth. The body of the uterus was rounder than natural from the os tincæ to the fundus; the muscular structure was thicker than ordinary. There was no physiological hyperæmia, no flocculent deposit, nor any decidua. The interior of the body of the uterus was paler than usual, except at the lower part of the cervix, which was congested in small vascular spots near the os tincæ. The broad ligaments were slightly congested and were carefully removed to obtain a distinct view of the ovaries. The left Fallopian tube was very slightly attached to the ovary, by a small portion of the fimbriæ, and was not very vascular. The tube was cut off near the uterus, and this was not more congested than the interior of the uterus. The ovary was very much enlarged, of a deep red appearance, especially at the uterine end. The small white spot of about two lines in length, indicates where the rupture occurred, and from which two or three pints of blood issued. The ovary measured one inch ten lines in length and one inch four lines in breadth. The distal end was more prominent and fuller than the uterine, but it was not as deeply congested. The epithelial covering was thinner, and gave no evidence of laceration. This tunic, as well as the tunica propria, was divided and turned back displaying the Graafian follicle, with its clear and transparent pearl-colored membrane, through which a small solid body could be seen. It appeared to be directly in apposition with the small circular white

TEXT FOR PLATE V.—1. Cavity of uterus, no decidua. 2. Cavity of cervix, congested. 3. Left ovary, highly vascular, and spot where rupture occurred. 4. Ovarian epithelium, and tunica propria cut open showing. 5. Enlarged Graafian follicle. 6. Thin transparent tunica pellucida or vitelline membrane, with a small body size of a pea attached. 7. Right ovary, very large. 8. The remains of three cicatrices. 9. Ovarian epithelium, dissected off.

spot externally, in the region of the vitellus. The follicle measured thirteen lines in length, and ten lines in breadth. The Graafian follicle was attached posteriorly to the ovary in one-third of its superficies, the rest was free. The right ovary was enlarged, the Fallopian tube was free and moderately congested at its extremity. The epithelial coat was divided and rolled over showing its slight lobulated appearance with three cicatricial points of rupture.

Remarks.—Considering the number of these cases recorded, they cannot be considered so very uncommon, or as rare as abdominal pregnancy. Usually cases of extra-uterine fœtation are tubal. Velpeau accepted the opinions of Blainville and Serres, who assisted him in dissections and investigations of the four supposed cases of ovarian pregnancy, and he states that "he had evidently been imposed upon in this matter." The fourth case, however, there was great difficulty in deciding upon, after having carefully isolated the Fallopian tube. The débris of the conception was contained in a sac lying between the peritoneum and the tunica propria of the ovary. In truth, the impregnation was *on* the surface of the germiniferous gland outside but *not* in it. Velpeau does not even, as he says, pretend that the ovum has never been observed on the surface of the ovary, but that when once verified, " it has never yet been found enclosed in the envelopes of the organ as in a cyst."

We are all aware that the unimpregnated ovum is believed to leave the ripe Graafian vesicle in order to enter the fimbriated extremity of the tube, and descend into the uterus, in either of which it may become fecundated, being discharged either at the beginning of menstruation, after, or during its flow. The epithelial tunic at present is considered as nearly allied to the mucous coat of the tube, though without its cilia, and bathed in a peculiar liquid. Sometimes only a small part of the fimbria is in connection with the ovary, and the ovum may pass along some of its grooves or furrows to the tube itself. It is even considered as not necessary for any part of the fimbriated extremity to be in close apposition. These physiological laws are so amply verified, that it naturally suggests the question whether the ovum could not be impregnated before leaving the follicle, and establish a pregnancy in that organ, enclosed, "as in a cyst."

The ovum may during the act of connection be in the stage of development, and preparing to go through the process of delivery, when it is met by the spermatozoids and conception follow ; or it may be expelled from its cavity into the tube, or drop into the abdominal cavity : or it may be arrested on its way of delivery and before its escape, and become an *ovarian gestation* dwelling in the ovary itself " as in a cyst."

It is not necessary, that the impregnated ovum should receive its nourishment or be attached to mucous surfaces. The germ may attach itself and live till it has completed its full term of intra-uterine existence in the abdominal cavity, without having any connection with the uterus, tube or ovaries. The ovum is nourished by endosmotic action.

From the history of other cases reported at as early a period of pregnancy as my own (Kammerer, N. Y. Med. Journal, 1865 ; I. G. Porter, Amer. Journal Med. Sciences, 1853), there can be no doubt of the embryo being enclosed in the envelopes of the germiniferous gland. Velpeau admits his was on the out-side and between the peritoneal coat, or, more correctly, the epithelial tunic, and the tunica propria, and is an exceedingly rare place for gestation to be recognized. But considered as an impregnation, and covered by a peritoneal coat, the spermatozoids must have penetrated that membrane, and it lends confirmation to the facts advanced in these remarks.

REPORT OF A CASE

OF

PROGRESSIVE FACIAL ATROPHY.

BY E. C. SEGUIN, M.D.,

Clinical Professor of Diseases of the Mind and Nervous System, College of Physicians and Surgeons, New York.

Progressive facial atrophy is such a rare disease that every case is worthy of being recorded.

CASE.—Delia H., aged ten years, was brought to my clinic for diseases of the nervous system, at the College of Physicians and Surgeons. The mother gave the following brief history : Delia had enjoyed good health as a child. At the age of five years there was noticed a greenish spot, "like a ringworm," on the left cheek, midway between the malar bone and the angle of the mouth. There was no pain. In the course of one year, the cheek became depressed and the mouth crooked, being drawn upward and to the left ; no other spots were noticed. The disease has steadily progressed, without pain. General health and mental development have been satisfactory.

Examination.—The child is well grown and healthy-looking, but the left side of her face is disfigured by atrophy of a large part of the cheek. The angle of the mouth is drawn upward and to the left ; the upper red border of the lip on the left side is small, almost linear. The region between the mouth and the malar eminence is the seat of a depression capable of receiving half an English walnut. At a distance the skin appears healthy ; but closer examination reveals the following points : a whitish spot exists in the centre of the atrophied region, about the size of a five-cent silver coin. The rest of the skin of the head, face and neck is normal. The deeper tissues of the cheek are much wasted, and the thickness of the cheek is less (one-third) than that of the opposite side. The very thin left upper lip, and the left cheek are not at all adherent to the subjacent bone. The rest of the left face is rather thin, but

FIG. 25.

presents no positive atrophy. The tongue and soft palate are normal. The left superior maxilla is, however, much atrophied ; its alveolar process not being more than one-half as thick as that of the opposite side. The teeth are more regular on the diseased than on the healthy side ; they are of good size and fairly preserved ; the space between the two upper middle incisors is abnormally great : there is no syphilitic or rachitic peculiarity about them. The anterior portion of the hard palate is

deeper on the left side. The lower jaw is normal. The pupils are equal and normal. Motion is well performed in the atrophied left cheek, as well as in other parts.

Sensibility.—Esthesiometer points are differentiated on the forehead, of 15 mm. on either side, on cheeks at 11 or 12 mm. on either side, on the lower cheek and over jaw at from 6 to 10 mm., on the red surface of the lips 2 mm., equally well on both sides. Pricking is well and similarly felt on each side. The electrical reactions were carefully studied. First, to *Faradism.* Nervous reaction (Kidder's induction machine), on right and left substantially equal, with minor current from posts A. B. and four inches of cylinder withdrawn, contractions obtained. Muscular reactions: one sponge on malar eminences and the other electrode on the lip or cheek, externally and internally, cause similar contractions on either side. Every muscle of the lip and cheek on the left side is present, and responds normally. Second, to *Galvanism.* With 8-9 elements reactions of nerves and muscles are normal; on both sides jerky contractions and C O C C ≻ A N C C.

Vasomotor phenomena.—Except the whitish spot above described, no difference is to be noted between the cheeks; when the patient blushes both sides of her face and head are equally suffused, and the same is true of the effect of nitrite of amyl. The mother of the patient thinks that she does not perspire in the wasted region.

A treatment by galvanism, cathode placed strobile on the wasted region for five minutes with a medium current was faithfully carried out for several months without result. The patient was not again seen until January 14 of the present year, when the following notes were taken.

Patient is large and well-developed, has menstruated normally for a few months.

The left cheek presents very much the same appearance, except that the spot of discoloration is no longer visible. The atrophy lies between the left malar eminence and the mouth, and involves the subcutaneous fissures of the cheek and left upper lip, and the left upper maxilla, in its palatal and alveolar parts. The wasting, both in bone and lip, reaches just to the median line. Can purse lips and draw mouth upward and backward well; the muscles of the atrophied region acting as well as those of the healthy side.

Sensibility.—To the lightest finger-touching test patient thinks that she feels less acutely in the left infra-orbital and mental regions than on the right, but the esthesiometer reveals no difference—both sides are normally sensitive. States that when she washes her face she feels the coldness of the water equally on both sides.

Faradic and *Galvanic* tests were carefully made, and showed no qualitative change in nervous and muscular reactions. To *Faradism* with a metallic electrode within the cheek, the muscles within the atrophied region contracted more quickly and fully than those of the healthy side (with the same weak current). This was probably due to diminished resistance in the affected cheek by the disappearance of the skin and some areolar tissue. The patient's mother states that she now blushes equally on both sides of the face. I did not feel justified in harpooning the face for muscular fibres, hence can make no statement upon their condition.

One case has been published by Prof. W. A. Hammond. The patient was shown to the New York Neurological Society, and Dr. Hammond stated that the muscular fibres of the atrophied region were smaller than those from corresponding muscles on the healthy side; but showed no degenerative changes.

PLASTIC OPERATIONS FOR LOSS OF NOSE, LOWER EYELIDS &c
OF THOMAS SABINE

PLASTIC OPERATIONS

FOR LOSS OF NOSE, LOWER EYELIDS, ETC.

BY THOMAS T. SABINE, M.D.,

Professor of Anatomy, College of Physicians and Surgeons, New York.

Thomas Colt, age 25, U. S.

Patient entered Bellevue Hospital in 1871, with so-called lupoid (?) ulceration of face. This finally healed after having destroyed the nose, both lower eyelids, part of the lips, especially the upper, and part of cheeks. He came under my notice in 1878, at which time he presented the appearance seen in the two upper photographs of plate VI. The corneæ, from constant exposure, had become hazy, so that he was unable to read, and the mouth could be closed but little more than is seen in plate VI. I felt disinclined to do any thing, as the tissues to be operated on were cicatricial, and I feared that the old disease might return. I heard nothing from him for some months, when he again applied to me. I consented to do an operation for restoration of the right lower lid, intending to be guided by the result of this as to any further proceedings.

May 28, 1878.—Fig. 28.—An incision ABCD, was made, marking out a flap which was dissected up as far as AD. The skin was then removed from the semicircle a, and the flap doubled upon itself along the dotted line and raised in such a way that the surface above the dotted line looked backwards toward the eyeball, the surface below looked forward, the part at dotted line forming margin of the lid; b fitted into a; which part of the

FIG. 28.

[37]

operation was done for the purpose of holding the lid in place. By thus raising the flap, a raw surface was left which had been occupied by the part of the flap below the dotted line. To fill this, and to hold up the lid, an incision BEFG was made and the flap dissected up. It was then turned at a right angle and stitched to the margin of the space to be filled, BE coming to BC (which had been raised to height of dotted line), EF to lower part of DC, GF to cheek part (as distinguished from flap part) of BC.

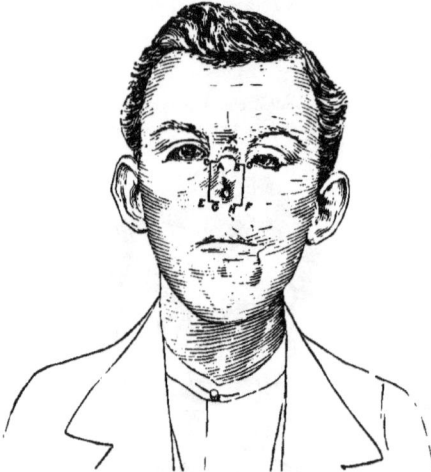

By dragging the flap BEFG into its new position, the raw surface, which would have been left at the part from which it was taken, was nearly closed, the portion of the flap G having been drawn to B, and the closure was completed by stitching GF to BE. As this left the skin of the cheek below FE puckered, the triangle FHE was removed and FH stitched to EH.

The result of this operation was so good that a nearly similar one was done on the left side.

October 3, 1878.—After these operations the corneal haziness nearly disappeared, and the patient was able to read the finest print.

December 19, 1878.—Operation for relief of deformity of mouth, Fig. 28. An incision passing entirely through the upper lip was made from A to B. This flap was dragged down so that A could be brought to E, leaving a triangular space where AB had been. Incisions EACD were made and all the tissue between AE and margin of left side of lip removed. The point A of the lip flap was then stitched to the angle of the mouth E, and the triangular flap ECD, which had been dissected up, fitted into the triangle left by the pulling down of lip flap, C coming to B, DC to upper edge of AB and EC to lower edge of AB. The raising of flap ECD of course left a raw surface at this part which was easily closed by stitching upper part of CD to CA.

June 20, 1879. Operations to relieve deformity of inner canthi. The overhanging skin at each canthus was slit toward but not as far as median line, the incision dividing also the conjunctiva on the under surface of fold and the tissues between. The edges of the incision then separated, thus enlarging the palpebral fissure. Along the edges the skin was stitched to conjunctiva.

November 1, 1879.—The patient being very solicitous to have a nose made, I explained to him that the usual operation of taking a flap from neighboring or distant parts would be useless as there was no bony frame-work to hold it in position, and that it would consequently shrivel to a mere knob of skin. Knowing of a case reported by Hardie, of Manchester, England, in which he had made use of the last phalanx of the left forefinger to form a nose, the idea having been suggested to him by his house-surgeon, Mr. Tytler, I offered to do a somewhat similar operation if my patient would assume his share of the risk, which was that the operation might fail, and he would then have lost a finger and still have no nose. He expressed willingness to lose a second finger if the first operation should fail.

The last phalanx of middle finger of left hand having been frozen, the nail was torn out and the matrix scraped and burned with nitric acid.

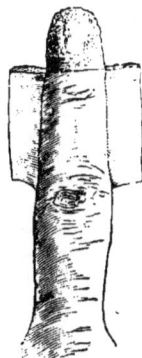

FIG. 30.

December 12, 1879.—The surface from which the nail was torn has healed, the matrix having been apparently destroyed.

A plaster of Paris jacket was fitted to the chest, and when hard was slit up at one side, so that if difficulty of respiration should occur during the operation, it could easily be removed. This was joined to a plaster of Paris helmet, which covered the upper surface and sides of head, and back of neck.

A curved incision, AB, Fig. 29, was then made, and the skin and fascia detached from parts beneath, so as to form a pocket represented by the dotted outline. Incisions BDFH and ACEG were then made, the edge of knife being directed deeply toward the middle line, to enable the edges of the two flaps to be turned toward the median line in such a way that the under raw surfaces of the flaps would look nearly forwards. The skin was then so dissected off the flaps as to present two raw surfaces. Figs. 31 and 32.

From the lower half of last phalanx the entire skin was removed. Fig. 30. An incision was made along the middle of the palmar surface of the second and upper half of the third phalanx, and two transverse, one opposite the palmar surface of the first phalangeal articulation, the other opposite the middle of the last phalanx. These transverse incisions passed about one-half round the finger. The finger flaps thus formed were dissected up as deeply as the sheath of the tendons. A silver suture, armed with a needle at each end, was then passed transversely through finger near the

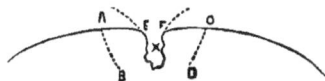

FIG. 31.

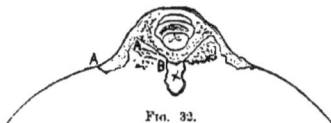

FIG. 32.

tip. The needles were passed into the pocket, brought out at X, and traction being made on the wire, the finger tip was drawn up into the pocket, in which it was concealed. The ends of the wire were then twisted together, holding the finger in place. The edges of the finger flaps were then stitched to BDFH and ACEG the under surface of the face flaps coming against the under surface of the finger flaps, and the denuded skin surface of the face flap coming against the raw side and palmar surface of finger. Fig. 29. After the finger had been partly stitched to the face, the patient, who had become more and more cyanosed, stopped breathing. A laryngotomy was done, and artificial respiration made, so that he soon breathed easily. The stitching of the flaps was then finished.

In Fig. 30 is seen the finger, ready to be placed in position, with the lower half of the last phalanx denuded of skin, and the flaps thrown outward, ready for application to the face.

Fig. 31 shows a transverse horizontal section of part of the face. X indicates the nasal opening (it should not be closed at the lower part, as it communicated with the naso-pharyngeal space). AB and CD mark the incisions backward and inward toward the median line, making flaps on each side, BAE and DCF which were turned inward. The skin from surfaces AE and FC was dissected off, so that each flap had two raw surfaces, AB and AE and CD and CF.

In Fig. 32 can be seen how the finger flaps and face flaps came together. AA'B is the line of junction of the under surface of finger flap and raw surface of side of finger with the face flap. A'B being the part of face flap from which the skin had been removed, and which, being turned somewhat inwards, came in contact with the side of the finger, and not with the under surface of the finger flap. The hand and arm were then fastened to the jacket and helmet by plasters and bandages.

The patient was fed through a tube introduced beneath the hand at the right corner of the mouth, until the finger was amputated. The jacket and helmet proving irksome were removed on the sixth day, and the hand was afterwards held in place by adhesive plaster and roller bandages.

As the finger was nourished through the digital arteries, and probably but little through the adhesion between the face and finger, it was thought necessary to compel a proper supply between the face and the finger. On January 3, 1880, an elastic ligature was passed round the finger opposite the lower part of the first phalanx. On account of pain this had to be removed in a few hours.

Jan. 5, 1880. The end of the finger has not united to the pocket. From retraction of the pocket flap and slipping down of finger the tip is exposed to view.

Jan. 8. A needle armed with a silver wire was passed deeply through the right side of the finger beneath the digital artery, and the two ends of the wire attached to a Wood's varicocele clamp.

Jan. 16. Wire removed. *Jan.* 22. Left digital artery secured. *Jan.* 25. Left wire removed.

Jan. 27. An incision was made down to the bone on the sides and palmar surface of finger. After this had been done the finger became pale, and the amputation was postponed.

Jan. 30. With the assistance of Dr. Sayre the finger was amputated, with knife and bone forceps, at the middle of the first phalanx. The stump of the finger attached to face bled freely, showing that the circulation was good. The part of the first phalanx attached to the finger was bent at right angles, so that the amputated surface looked backward, the part of the first phalanx attached to finger thus forming a septum for the nose, the first joint forming the tip.

Oct. 8, 1880. The finger tip now lies in front of the space between the inner canthi. The matrix was not destroyed as was supposed, as the nail has partly reappeared. It was softened with caustic potash, removed with forceps, and the matrix scraped.

Dec. 28. The skin over the site of the matrix was removed, and the tip of the last phalanx cut off with bone forceps. A small flap was then slid down, and fastened on the denuded finger tip.

Since the above note several minor operations have been done which it is not necessary to mention in detail. At the present time the finger has sunk downward a little, so that the lower eyelids are somewhat dragged upon. The amputation of finger should have been done nearer the hand, as the tip of the nose is not prominent enough, partly from atrophy of that part of the first phalanx attached to finger, and partly from the sinking backward and atrophy of the finger itself.

DUPUYTREN'S CONTRACTION OF THE FINGERS.

BY ROBERT ABBE, M.D.

Surgeon to the Out-Patient Department of the New York Hospital; Consulting Physician to the Hospital for the Ruptured and Crippled.

In 1832 Dupuytren distinguished, among the many deformities and contractions of the hand, one that was characterized by a drawing down of one or more fingers toward the palm, by a firm band extending subcutaneously toward the wrist. It developed slowly, was not preceded by cellulitis, nor by injury, and came almost exclusively in men usually past middle life.

It is four years since Mr. Adams of London again drew attention to the malady, and proposed for its relief the simple and perfect expedient of subcutaneous division of the contracted band. British Med. Journal, June 29, 1878. Dupuytren and others had advised open operations, which were not always successful and often were disastrous.

The case I am about to relate is a perfect type of this rare trouble, and presents one or two features of special interest. The patient, rather a large man, aged forty-five years, came to me June 9, 1881. He had been a cloth cutter for fifteen years. He had never had syphilis, rheumatism or gout. His occupation required the use of large scissors in his right hand but nothing in his left.

Seven years ago the left little finger began to draw down toward the palm without known provocation, and had grown progressively worse, until it rested flat on the palm though it could be raised by a little force to the extent shown in Fig. 33.

One year ago the ring finger of the right hand began to contract in the same fashion and had progressively become worse until it was as shown in Fig. 34. Of late it became such an inconvenience and so painful that it prevented his working. To all appearance there was a tendinous band as of the flexor tendon rising from its bed, stretched tightly beneath the skin of the palm from the base of the second phalanx of the affected finger in each hand, well up towards the wrist. The skin along this band was puckered and intimately bound to it. The joints of the affected fingers were apparently not anchylosed, but extension beyond the amount shown in the figures was prevented by the band referred to. Careful study made it evident that these were each a

[41]

fasciculus of the palmar aponeurosis tightly contracted, intimately related to the skin in parts, and lying above the flexor tendon, which was not diseased.

On June 9th I resorted to Adams's operation on the little finger of the left hand (Fig. 33), dividing the band at two points in the palm near the root of the finger, where it was most tense and accessible. I did this without ether, and on account of the pain the patient refused to let me make a third division of a *lateral* band, on the outside, which prevented complete extension by about thirty degrees only. The greater part of the deformity, however, had been repaired by the two cuts. The finger was bound in extension upon a posterior padded splint. Five days later it was found capable of flexion and extension through the limits gained by the operation. But another unexpected result was reported of which the patient volunteered this statement. For two months the hand in the right palm had been so constantly in pain that he could not move this hand up to touch his head without almost unbearable suffering, and at times, especially if the palm was struck, it was very painful. *Directly after the operation on the left hand the right ceased to hurt* and he has found that he can now lie at night with his right hand behind his head and sleep on it without pain, which he has not been able to do for months.

On June 21st I operated under ether, making subcutaneous division of the band in the right palm at five points half an inch apart. Its adhesion to the skin made this number necessary. The finger was well released, straightened out parallel with the others, and subsequently regained suppleness and freedom. The joints and tendons were intact. The patient has been watched for half a year and no recontraction has taken place. He resumed his work of cloth cutting and has now no pain or return of trouble. In cutting I used the small tenotome of the ophthalmic surgeons, and much prefer it.

Fig. 33.

Fig. 34.

It has been proved that the mechanical extension devices fail to overcome this contraction, and that there is no tendency to recontract after operation as there is in burn cicatrices. Adams believes this malady can almost always be traced to one constitutional origin, namely, gout. Bryant thinks that from its occasional symmetry it has probably a constitutional and not a local origin (injury). From the clews derived from the history of this case, I am inclined to think there may be a central nervous origin. The right hand became affected after the left had been progressing several years, and was most curiously relieved of all pain by operation on the left, giving strong suspicion of a reflex relation between the two. I may say, however, that there were otherwise no ataxic symptoms.

THE PATHOLOGICAL ANATOMY

OF A CASE OF

SPINAL CARIES WITH PARAPLEGIA.

BY V. P. GIBNEY, M.D.,

Of the Hospital for the Ruptured and Crippled, New York City.

The accompanying plate represents the appearances found post-mortem in a male child æt. six years, the clinical notes of whose case have been presented to the New York Pathological Society. Yet in view of the reproduction of the sketch made at the time of the autopsy, I propose to briefly give the more important facts in the case, that its pathology may be the better understood.

The boy had marked deformity of the spine from caries of the vertebræ of eighteen months, and paraplegia of four months' standing, when he came into hospital, August 18, 1874. The paraplegia was marked by the characteristics of Pott's disease: viz.: complete loss of voluntary power, very little atrophy, reflexes exaggerated even to the exhibition of spinal epilepsy, &c. Symptoms of profound myelitis were developed, bed-sores forming over the nates and thighs; pulmonary hyperæmia produced, at times, very alarming symptoms, and he finally died, thirteen months after admission to hospital, seventeen months after the invasion of the paraplegia, and two years and seven months from the inception of the disease.

The treatment consisted of mechanical support for the head and spine, counter-irritation, massage, tonics, &c. Dr. Janeway assisted me at the autopsy. The angular deformity of the spine extended from the first to the twelfth dorsal vertebræ, the apex being at the seventh. The various organs presented nothing worthy of note in this connection. Within the spinal concavity lay a tumor, two inches high and three and a half inches broad. This was an abscess, the contents of which had degenerated into a white, cheesy mass, well shown in the left hand figure of the plate. Its walls were formed by the anterior common ligaments, thickened connective tissue and the pleura. On vertical section no trace of the the body of the eighth dorsal vertebra could be found, and only small portions of the bodies of the sixth, seventh and ninth. Pultaceous matter occupied the place of the body of the eighth vertebra, and pressed on the spinal cord, which at this point was anæmic to a high degree, yellow, and very small. The absence of blood vessels in this compressed portion was notable. The greater portion of the cord was removed, and the sketch made without delay. The cord was placed in Mueller's fluid, and was well preserved eighteen months later, when Dr. Seguin kindly made a microscropical examination, cutting specimens from 1, the cervical enlargement, 2, the upper dorsal region four-fifths of an inch above the limit of compression, 3, the same distance below the lower limit of compression, and 4, from the lower part of the lumbar enlargement.

"The above sections were examined in two ways: 1. In a saturated solution of acetate of potassa, without staining. This was to show the granular bodies. 2. By Clarke's method, to show the atrophy of nerve tube in parts, and the sclerosis of the neuroglia. Sections No. 1, treated by acetate of potassa, showed very exquisitely the lesion of ascending degeneration. The columns of Goll, or posterior median columns, were filled with granular bodies. Sections No. 2, seemed quite extensively altered, granular bodies being found almost throughout the section—probably from pressure causing ischæmia of the parts. Besides the above, granular bodies were found in small number in the external part of the posterior lateral columns—the ascending cerebellar fasciculi of Flechsig.

"Below the seat of pressure, sections 3 and 4 showed the usual descending degenerative changes in the white columns, the mass of granular bodies occupying the outer and posterior part of the anterior lateral columns. The sections prepared by staining with carmine and by Clarke's method afterward, showed the same ascending and descending degeneration, evidenced not by granular bodies, but by atrophy of nerve fibres and increase of the neuroglia. No lesion existed in the gray matter, and the cells of the anterior horns seemed normal. In other words, this examination shows that except at the seat of pressure, there were no lesions other than those of ascending and descending degeneration."

This case illustrates the now accepted theory of the production of paraplegia, and the different tissues successively attacked. First we have the carious ostitis of the bodies of the vertebræ, then by contiguity, an extension of the morbid process of the perimeningeal areolar tissue, the vertebral ligaments, and the dura mater. This gives us the "*pachymeningite externe*," which is characterized by vegetations on the external surface of the dura mater, and by consecutive carious alterations which blend with the carious detritus in the bone. The cord, consequently, becomes compressed, a focus of myelitis is induced, and from this focus a tranverse myelitis arises. This is followed by a fasciculated sclerosis ascending through the posterior and descending through the lateral and anterior columns. It is not necessary to have any angular deformity; the hospital records containing the history of several cases wherein complete paraplegia existed, and yet no bony deformity was found. Such cases are, however, exceptional. Pressure is often removed by the formation of an abscess, and recovery from the paralysis ensues, sometimes quite unexpectedly. Subsequent restoration of motion, in cases like this one, is the result of re-establishment of the circulation in that part by the removal of pressure. No fact is so well established as this one, viz.: that cases of paraplegia, with evidence of lesions as grave as those I have mentioned, do recover. "It must be remembered that the sclerosis involves the neuroglia, and that the secondary degeneration takes the place of the white fasciculi connecting the nerve tubes. Even should the envelope of myeline be destroyed, impulses can be transmitted through the axis cylinder, and should this be destroyed, the degeneration existing in tracts or bundles, other nerve tubes which remain intact may serve for the transmission of impulses."[*]

* "The Paralysis of Pott's Disease," Journal of Nervous and Mental Disease, April, 1878.

THE HISTORY OF THREE CASES

HIP-DISEASE IN THE THIRD STAGE.

BY A. B. JUDSON, M.D.,

Orthopedic Surgeon to the Out-Patient Department of the New York Hospital.

CASE I.—Was of a boy, æt. six years, who presented an enormous abscess and all the usual symptoms of the third stage of hip-disease, which was of nineteen months' duration. The abscess had advanced so far that a spontaneous opening occurred on the same day in which the patient was first examined, and before treatment could be instituted. The child's general condition was bad. The limb was strongly flexed and adducted. The slightest attempts at motion elicited screams of pain. Exsection had already been urged by a medical attendant.

Mechanical treatment was begun the sixth day after the patient was first examined. The apparatus used was the long hip-splint, first described by Dr. C. Fayette Taylor in 1867. Its essential parts are a pelvic band carrying two perineal straps which are applied to the ischiatic and pubic regions of the pelvis, for counter-extension, and a strong upright containing a sliding bar moved by a rack and pinion and having a rectangular piece extending under the sole of the foot. Adhesive plasters are attached to the limb and buckled to the foot-piece of the splint to ensure extension. A piece of webbing was buckled round the splint and the lower part of the thigh, although it is believed that the fixation power of the apparatus is increased by the substitution for this webbing, of a U-shaped piece of steel which retains the femur more nearly in a line parallel with the upright of the splint. This apparatus was used with a two-fold object: first, to afford a reasonable degree of immobility to the joint, and, secondly, to facilitate locomotion by acting as an ischiatic crutch. With the addition of an elevated shoe to the foot of the unaffected limb, the patient was about the house daily from almost the very beginning of treatment, and the affected limb was as free from concussion in locomotion as if it had been a naturally pendent member, calling to mind the words of the entertaining writer, M. Hennequin: "Mais le corps humain peut il conserver pendant des mois entiers l'attitude verticale, touchant le sol par un pied seulement ? Evidemment non, cest au-dessus de ses forces. L'avenir nous réserve sans doute de grandes surprises, et ce qui est impossible aujourd'hui deviendra peut être facile demain." (Archives générales de médecine, Jan. 1869, p. 64.)

The first application of a splint to a patient in the third stage of hip-disease is a matter of some difficulty. The apparatus is constructed as if it were to be applied to a symmetrical figure; hence when first brought near the patient the symmetry of the splint throws the deformity into such marked relief as to make it seem impossible to use the apparatus. The free ends of the pelvic band may extend obliquely upwards over the thorax in front and behind, on account of the extreme adduction. The perineal straps may be far from occupying the places which they would fill if the patient's body were symmetrical, indeed, it may be impossible at first to use them on the affected side. But with care and gentleness the instrument can be so arranged as to permit of a slight amount of extension and counter-extension. This is attended inevitably by a partial but most grateful arrest of motion and is followed immediately by a gradual reduction of the deformity. In a few days the symmetry of the patient's figure will be so far restored that the splint is properly and comfortably worn. The pelvic band can then be lowered to its place below the level of the anterior superior spine of the ilium, the perineal straps will adapt themselves to the ischiatic and pubic regions, the flexion of the femur will be materially diminished, and the adduction will have given place to abduction. This new abduction, with the consequent apparent lengthening, may become so great as to cause anxiety for the ultimate position of the

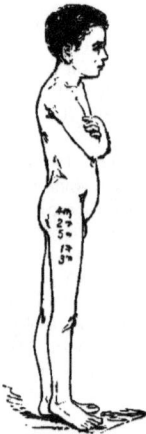

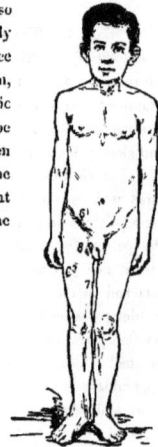

FIG. 37.   FIG. 39.   FIG. 38.

limb. This, however, disappears in time. As the patient gathers strength from the absence of pain and the return of sleep and appetite, locomotion without crutches will be resumed, and it will be seen that the fixation afforded by the splint is so well adapted to the requirements of the case as to obviate pain and promote the reparative processes and yet not so firm as to prevent the gradual disappearance of the abduction and the further diminution of the flexion in obedience to the unconscious efforts of the patient to put the limb in the most favorable position for locomotion. These views of the action of the hip-splint in the reduction of the deformity of acute hip-disease are at variance with much that has been written on this subject, but they are founded on clinical observation.

The changes from adduction to abduction and finally to a symmetrical position of the limb were observed in due order in this case. With the hip-joint thus protected from undue motion and also from pressure and concussion, the patient was enabled to pursue the ordinary occupations of a boy of his age while the reparative process gradually supplanted the ravages of the disease. Recovery, however, was not immediate. The abscess already referred to was followed at irregular intervals by other purulent collections which were incised or opened spontaneously until nine sinuses were established about the joint, all leading to carious bone. Five of these sinuses extended in a line down the outer side of the thigh from the trochanter to near the middle of the shaft of the femur, as seen in the cut, Fig. 37. The position and arrangement of these sinuses from one of which a fragment of cancellated bone was extended, the nature of the discharge, which was frequently offensive, and the character of the resulting cicatrices show that this was a case in which the shaft of the femur was to a considerable extent involved in the destructive osteitis. Although the progress of the case was generally towards recovery, there were stages in which the general condition of the patient was seriously affected. On such occasions the appetite failed, the tongue became coated, lassitude and irritability supervened and frequent ephemerae indicated how profoundly the system was affected by the local disturbance. At such times and more especially throughout the early and more critical period of the disease, cod-liver oil and other roborants were freely prescribed. The fact that the patient was enabled to move about and to amuse himself in the open air and sunshine was believed to be especially useful in supporting his general health and thus indirectly promoting the recovery of the involved joint.

When the improvement in his general condition, the tolerance of motion in the joint and the disposition of the sinuses to close indicated the propriety of gradually relaxing the treatment, the splint was worn for some time with only a slight amount of traction, and finally the adhesive plasters were removed and the splint was worn for several months suspended merely by webbing passing over the shoulders, making, in fact, an ischiatic crutch. An elevated shoe on the foot of the unaffected side enabled him to walk briskly and at the same time to regain whatever motion could be got from a joint so thoroughly disorganized without exposing the new tissues to the violent concussion inseparable from ordinary locomotion.

The patient was under treatment two years and five months. His present condition, six months after treatment, is shown in the wood-cuts, Figs. 37 and 38. It is extremely favorable in view of the extensive destruction which had occurred in the joint, and the prolonged strain to which his system has been subjected by the disease. The limb is in good position, neither abducted nor adducted, and flexed at a slight angle, sufficient to allow him to sit comfortably, and yet not enough to interfere with locomotion. The motions of the knee are perfect. He walks with firmness, runs rapidly, and never uses a cane. The limp which accompanies rapid motion, and is slightly perceptible when he moves slowly, is partly the result of an inch of shortening and partly due to the absence of motion in the joint. That shortening comes not so much from the loss of bone at the upper extremity of the femur, as from a disparity in the size of the bony structures of the two legs, is illustrated in the outlines of the feet, Fig. 39, a disparity arising not only from disease and desuetude of the affected limb, but also, perhaps, from over-use of the unaffected limb. The remarkable locomotive power possessed by the patient in view of the size and importance of the joint affected, illustrates the ease with which motion is transferred from an impaired joint to the lumbar region of the spinal column and the hip-joint of the unaffected side. The auxiliary motion of these parts acquired thus in youth may

[47]

be expected to increase with the further growth of the patient. The position of the sinuses is shown in the figures. The cicatrices are firm, deeply depressed, and in some instances attached to the bone beneath. They are numbered in the order in which the sinuses made their appearance. The family history of this case shows no evidence of scrofula.

CASE II.—Was that of a girl, three years of age, whose mother died of consumption while the child was under my care. The family history showed that not only the mother, but also the maternal grandmother and three paternal uncles and aunts had died of phthisis pulmonalis. The disease was in the right hip, and had existed at least one year. Previous mechanical treatment had been by an immovable dressing of plaster of Paris, and afterwards by the use of a long hip-splint furnished

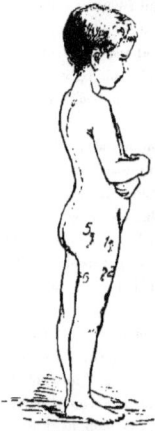

with a single perineal strap, and applied without adhesive plasters. This splint was constructed with a joint at the level of the knee, for the purpose of assisting locomotion, which was further facilitated by the use of Darrach's wheel crutch. When first seen the child presented the marked adduction and flexion of the thigh characteristic of the third stage, and had suffered for several weeks the intense pain which is usually the forerunner of abscesses communicating with the joint. The treatment adopted was identical with that of Case I. Under its use the pain abated, and the position of the limb improved,

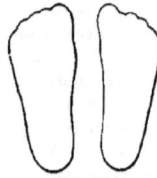

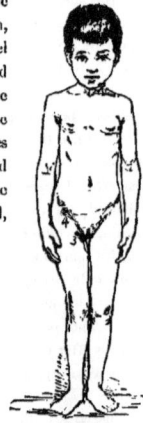

FIG. 40.   FIG. 42.   FIG. 41.

adduction giving place to abduction, and the flexion being materially diminished. But the abscess was not prevented. Five months after beginning treatment it was opened, and the sinus thus established on the outer surface of the thigh was followed, in the ensuing eighteen months, by five others, variously placed about the joint, which secreted an abundant and offensive pus, evidently from carious bone. The hip and upper part of the thigh were enormously swollen. During this period the treatment aimed at protecting the joint from motion and concussion and at fortifying the system so that Nature might check the destructive process, and substitute healthy or cicatricial tissue for that which was disintegrated. The treatment by tonics and roborants, viz., cod-liver oil, the more nutritive wines, chalybeates, &c., was apparently very much assisted by the use of a splint which allowed of locomotion in the erect position. With the exceptions to be mentioned, the patient, throughout

[48]

the entire treatment, was out of doors every day, walking with the aid solely of the ischiatic support furnished by the perineal straps of the hip-splint. There were occasions when, for several days together, it was impossible for the child to take her customary exercise on account of pain. At such times febrile reaction and emaciation threatened a fatal termination of the case by exhaustion, and these periods coincided with the development of new abscesses and sinuses. The most serious and prolonged relapse occurred when, from the death of the patient's mother, it became necessary, for a time, to entrust the mechanical treatment to the child's friends. Notwithstanding these complications, the discharge slowly diminished, the sinuses gradually closed, some degree of motion was restored in the joint, and the re-establishment of the patient's health showed that recovery was assured. Mechanical treatment was continued for two years and seven months. During the first half of that time strong traction was used, but during the latter half of the time, when it became apparent that fixation by the splint was no longer required, the apparatus was applied more loosely, and for several months it was worn only in the daytime, as an ischiatic crutch, to protect the new tissues of the affected part from pressure in standing, and concussion in locomotion.

The patient's present condition, eight months after the final removal of the splint, is well shown in the cuts, Figs. 40 and 41. The cicatrices are numbered in the order in which the sinuses appeared. Nos. 4 and 5 are attached to bone. The other scars are deeply depressed and attached to the fascia. Her health is perfect and she is able to walk and run without assistance of any kind. The position of the femur is favorable both for walking and sitting, there being no abduction or adduction, but a moderate degree of flexion, and the shortening is only one-fourth of an inch, evidently due to a diminution in all the measurements of the limb. The outlines of the feet are seen in Fig. 42. When she walks slowly it is difficult to perceive any limping, although the motions of the joint itself are so slight as to be of very little, if any advantage in locomotion. Fast walking and running develop a slight limp, but not enough to prevent her from participating in all the pastimes of her time of life.

CASE III.—The family history of this case is remarkably free from evidences of scrofula. This boy, when first examined, was seven years old and had suffered from disease of the right hip for four years. The patient's father, a surgical instrument maker, possessed unusual skill in the adaptation of apparatus, and hence the mechanical part of the treatment had not been neglected. The boy was provided with a long hip-splint which would have been serviceable had it not been constructed of such light materials that even a moderate degree of traction at the joint was impossible. At every step the instrument allowed the weight of the body to rest on the diseased joint. The usual signs of the third stage of hip-disease were present. Several weeks of severe pain had indicated the formation of an abscess, already recognizable by swelling, redness and heat on the inner surface of the thigh. Mechanical treatment was resorted to as soon as practicable in the same manner as in case I., which produced some relief from pain and a general improvement in the position of the limb. Suppuration progressed, however, until the hip and the upper part of the thigh were greatly distended and the pus was evacuated by incision or spontaneously, when four sinuses were established in the positions and in the order indicated in the cuts, Figs. 43 and 44, which presented the tumid and everted edges characteristic of sinuses leading to dead bone. The severity and persistence of the symptoms, the number and positions of the sinuses, the long continuance and often offensive nature of the discharge, and the character of the resulting cicatrices, of which Nos. 2 and 4 are attached to bone, clearly show that the case was one of destructive ostitis and disorganization of the joint. For many weeks the constitutional disturbance was severe. There was pallor with frequent hectic flushes and elevation of the temperature. Exacerbations of pain were partially relieved by the application of moist or dry heat. The diet was liberal and unrestricted in variety. Cod-liver oil and the ferruginous tonics were freely used. Notwithstanding the severity of the local symptoms and emaciation, the patient was usually able to be about the house, or out of doors, with the assistance of a pair of crutches, although it is probable that if he had not been previously dependent on them for a long time, he would have preferred to rely simply on the

ischiatic support furnished by the splint. The slightest attempt at motion in the joint was exquisitely painful, and the patient, soon after the beginning of treatment, perceived that locomotion and even movement of the body in bed were painless only when extension and counter-extension, with a reasonable degree of immobility, were enforced. At the end of a year it became evident from the diminution of the purulent discharge, the disposition of the sinuses to close, the tolerance of motion in the joint, and the improved condition of the patient that reparation was fairly established, and that fixation of the joint was no longer necessary. The splint was therefore removed and its

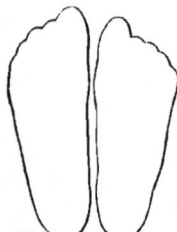

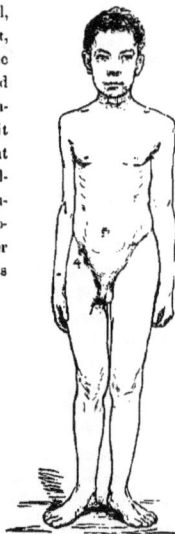

FIG. 43.   FIG. 45.   FIG. 44.

place was supplied by an instrument which, receiving the patient's weight on a single perineal strap, prevented his heel from reaching the ground and at the same time allowed of motion at the hip and knee. The crutches were then laid aside and this instrument was worn for three years, a longer time than was necessary, through excess of caution on the part of the patient's father, who assumed the subsequent care of the case.

The patient's present condition, eighteen months after all treatment was discontinued, is well depicted in Figs. 43 and 44. He is an active and robust school-boy, entering heartily into all the ordinary pursuits of a boy of his age. He takes long walks to and from school, and is a good skater.

When walking slowly there is no perceptible defect in gait. The limp which is developed in rapid movement does not prevent him from walking and running with great speed. He never uses a cane. There is half an inch shortening. The position of the limb is good, there being a moderate degree of flexion, enough to facilitate sitting, but not sufficient to interfere with locomotion. There is neither abduction nor adduction of the femur. Motion at the joint is practically abolished, so that his remarkable power of locomotion is due to vicarious mobility of the lumbar region of the spine, and the unaffected hip-joint. The dimensions of the affected limb fall below those of its fellow, as is seen by the outlines of the feet (Fig. 45), and in the fact that there is a difference of one-fourth of an inch in the transverse measurements of the patellæ.

In reviewing these cases it is evident that the favorable results cannot be attributed to the superficial or trivial character of the lesions. They were cases in which the principal indications for exsection were present. In one of them exsection was advised by one whose name is prominent in the history of this operation. Dr. Cheever, in the midst of the performance of what has been termed "the majestic and sanguinary hip-joint operation" (Medical and Surgical Reporter, Philadelphia, June 18, 1864, p. 383), remarked: "In this, as in every similar case, when you get into the joint you are surprised to see the amount of disease which did not appear externally" (Boston Medical and Surgical Journal, Aug. 22, 1878, p. 234). It may be inferred, therefore, that in the cases related, in which the external signs of disease lacked no element of severity, the lesions were destructive, invading the hard and soft parts of the joint, and were sufficiently serious to justify and even demand a resort to the most heroic measures. Operative procedure, however, gave place to mechanical treatment in accordance with views of pathology which may be stated briefly in these terms: the affection is not malignant, and is not seated in a vital organ. Sir Benjamin Brodie exclaimed: "Why should the disease be dangerous? The hip-joint is not a vital organ" (Clinical Lectures on Surgery, 1846, pp. 279, 280). Its activity depends largely on the motion of the part, and the pressure and concussion incident to its use in locomotion. This view and a reliance on the reparative power of Nature determined the adoption of a plan of treatment described above, which secured relief from acute pain, and which was followed not only by recovery, but by a degree of usefulness in the affected limb far beyond the usual results of exsection.

It has been claimed that exsection relieves the patient at once from the pain of progressive hip disease in the third stage. Mr. Hancock, in his elaborate argument for exsection, draws the following picture of a case of hip disease: "Look at a patient wasted to a shadow, confined to his bed, not for months only, but for five years, in constant pain and in the last stage of exhaustion from long-continued discharge, his hands employed night and day incessantly maintaining a fixed position of the limb, and endeavoring to prevent the intense agony which occurs on the slightest movement. Often have I seen the poor hip-joint patient, when all others have slept, still wakeful and anxiously engrossed with the one and monotonous task of steadying the knee and preventing movement. Look again at this patient when the operation is performed; his position now is no longer one of constraint and torture, it is one of comparative comfort and rest. He no longer suffers the extreme pain, he no longer exists in dread of the slightest movement or jar, his countenance loses its drawn and anxious appearance, the hectic subsides, and whatever may be the ultimate result, we at all events have the satisfaction of feeling that by the operation we have alleviated a very vast amount of suffering, almost beyond the power of endurance" (Lancet, June 1, 1872, p. 620). Great as is the relief thus depicted, it is not more marked than that which attends mechanical treatment in this stage of the disease. When fixation is secured, the anxious look gives way to an expression of repose, appetite and sleep return, and the reparative process begins.

As Mr. Hancock has suggested, the question of recovery after exsection is a momentous one. It is a serious question even when the operation is performed under the most favorable circumstances, such as surrounded the patients of Mr. Annandale. He was accustomed to operate at a very early stage of the disease. In one of his cases, a girl of sixteen years, the duration of the disease, previous to the operation, is recorded as three weeks. In the twenty-two cases which he reported in 1876, there were only five in which external sinuses existed. It will be seen from these facts that his cases were exceptionally favorable for operation, because they had not been weakened by exhausting discharges and long periods of suffering. Yet in these twenty-two cases death occurred in eight, at periods ranging from three to eighteen months after the operation (Edinburgh Medical Journal, February, 1876, p. 694).

As hip disease derives its desperate character (its *quasi* malignancy) from the difficulty experienced in securing rest, and not from the nature of the disease, which is sufficiently amenable to treatment when occurring in other parts of the body, it follows that the rate of mortality is diminished by providing efficient rest and avoiding the risks of operation.

Finally, in regard to the usefulness of the limb, the firm attitude shown in the figures,* the facility in walking and running possessed by the patients, and their ability to endure fatigue, leave but little to be desired for the results of treatment.

* The figures are from photographs in the library of the New York Hospital.

[52]

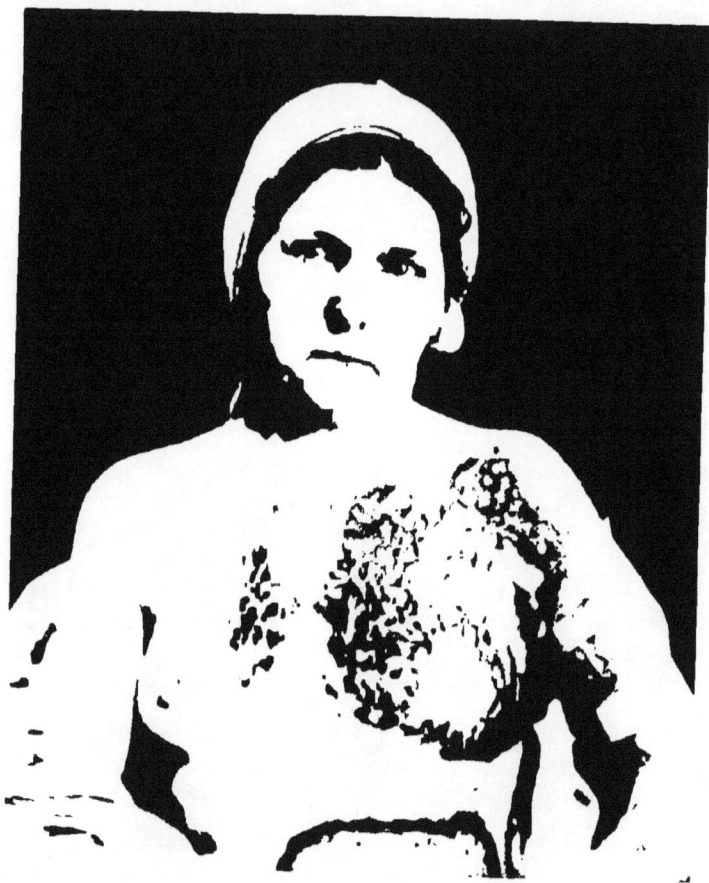

SKIN-GRAFTING IN THE TREATMENT OF A BURN.

BY GEO. A. VAN WAGENEN, M.D.,

Late House Surgeon Bellevue Hospital.

CASE.—Ellen Collins, æt. thirty-seven; Irish; married; was admitted to Bellevue Hospital, October, 1871, for a burn from kerosene oil. After five months' treatment the wound assumed the appearance seen in the illustration. It involved the whole right breast and arm, with a portion of the left breast; and had ceased to cicatrize. The nipple had been renewed and the wound of the left breast almost covered by grafts from her own person and those of her friends.

On March 10, 1872, I immersed an amputated leg in hot water and began grafting from it within three minutes after immersion. Thirty grafts were applied on the right side and covering a vertical space five by four and a-half inches, and almost all succeeded.

March 23.—Being unable to take the amputated limb immediately to my patient, I wrapped it in flannel and placed it behind a coil of steam pipes; beginning grafting *one hour and thirty-five minutes* after the operation. About one hundred grafts were applied, and on April 1, I counted eighty-nine successful ones.

On *April* 10, this photograph was taken to show the grafts of March 23. Another, taken three months later, showed the breast nearly healed, and the arm much improved. An acute pleurisy with effusion of the *left side* then set in and caused her death, on *June* 22, when everything promised success. A post-mortem showed that death was not directly due to the burn. She had received in all over 1500 grafts in less than a year; a large proportion of which were successful.

1. *History.*—The history of skin-grafting is not a long one. In 1847, Professor Frank Hamilton, of New York City, *suggested* transplanting pedunculated flaps of skin to heal large wounds; and applied this method in 1854.[*] But to Mons. Reverdin, of France, belongs the credit of first using detached grafts of skin. The successful transplantations were made October 16, 1869; and his paper on " Epidermic Grafting " was read before the Société de Chirurgie de Paris, in December, 1869.[†]

In May, 1870, Mr. George Pollock, of St. George's Hospital, London, began grafting in England.[‡] About the same time, Mr. D. Fides, of Aberdeen, Scotland, employed epidermic scrapings

* N. Y. Med. Gazette, Aug. 20, 1870.
† See Bulletin of same Society for 1869. Also Gaz. des hôpitaux, Jan. 11 and 22, 1870.
‡ For his cases and remarks see Transactions Clin. Soc. 1871.

(epidermic grafting). This method is also attributed to M. Marc Sée, of Paris. I do not know who has the credit of introducing it in America, but early in 1871, I saw my preceptor, Professor Henry B. Sands, apply several grafts, and he then spoke of it as something entirely new.

Mons. Ollier[*] claims priority over Reverdin; and says he grafted periosteum, seven years before, though his results were published three years after Reverdin's successful cases.

1. *Sources of supply.*—The skin from any part of the body, provided it is not too thick and horny, may be used. Formerly, I supposed I was the first to graft from amputated limbs,[†] but learned afterwards that it had been previously done in Boston.[‡] The results however, are unpublished.

I have grafted successfully from a limb which had not been kept warm, *eight hours* after amputation, and indeed it is said that in dead bodies the skin of animals and the lining membrane of an egg have been used successfully for this purpose.

3. *Method.* — Reverdin and Mr. Fides describe the cutaneous grafts as "epidermic grafting,"[¦] the former believing the epidermis to be the active part of the operation.[§] The graft usually includes both derma and epidermis; but should not contain any of the subjacent cellular tissue, because this prevents its attachment to the ulcer. A very ingenious combination of forceps and scissors for the removal of the skin has been proposed by Mr. McLeod.[**] My own method is, after shaving the hair, to take a fold of the skin between the thumb and finger, and cut a strip from its apex with a sharp, well-moistened scalpel similar to what is done in cutting microscopic sections. This strip should be made about a quarter of an inch wide, by two inches long, and should include the upper layers of the derma, with its rete mucosum. This is enough to make several grafts — which should be about one-eighth by one-quarter inch in size. Those first used were very small; and while grafts the size of the finger-nail will usually grow, better results are attained, if the same amount of tissue is distributed in smaller sections. The surface of the sore, which *must* be healthy and bright, should be thoroughly cleansed with moist lint until the mucus which covers it is removed; and the ulcer appears dry and raw. On this surface, without cutting or scraping, as was formerly done, the first row of grafts, their cut surfaces looking downwards, should be placed half an inch from the edge of the wound. A second row may be placed the same distance beyond the first. Thus arranged, both the graft and edges of the ulcer grow rapidly; and unite before the stimulated granulations rise and prevent healing, which they

---

[*] Gaz. Hebdomadaire, No. 13, 1872.
[†] For dates and success see history of case appended.
[‡] At Mass. General Hospital.
[¦] Bulletin of Surg. Soc. of Paris, 1869; also Gaz. des hôpitaux, Jan. 11 and 22, 1870.
[§] See wood-cut of microscopic section of graft appended.
[**] Bryant's Surgery, page 130.

do when the grafting is begun at the centre. It is not necessary to retain the grafts with plaster, as was first done; but cover them simply with lint, over which enough cerate has been spread to prevent adhesion. Retain this dressing in place with a bandage to prevent displacement for twenty-four hours, when it may be carefully removed, after thoroughly moistening it, so as not to detach the grafts. At the second dressing they will be firmly adherent. That the best arrangement of the new tissue is the one suggested, I have proved by taking one hundred grafts of different sizes from the same limb, placing them on the ulcer in squares—solid and hollow—circles, rings, rectangles, etc., and at different distances from the edge. With a healthy granulating surface, and grafts carefully applied, without blood or pus beneath them, nine in ten should "take."

4. *Growth.*—At the end of about twenty-four hours, the edges of the new tissue are bent down and adhere to the granulations beneath, while the centre is elevated. Within forty-eight hours the dead cuticle begins to separate and may soon be removed; but the deeper part of the graft is attached and its edges appear more and more sunken, until, at the end of the third, or fourth day, a depressed ring seems to surround it. This is darker than the granulations, about half a centimeter wide, and is really produced by the increased growth of the adjacent granulations. The graft has now nearly disappeared, or become transparent; if the depressed ring remains it has certainly "taken"; and will soon reappear as a spot of delicate cicatricial tissue, increasing in circumference as the dark ring enlarges. Its progress is first rapid; then becomes gradually slower, until it stops, as in ordinary cicatrization; but if a new graft is placed near by, its growth begins again. This growth is also most rapid in the direction of the edge of the ulcer, or an adjacent graft, both of which project to meet it, thus shutting off islands of granulations.

5. *Microscopic appearance.*—I have not examined the growing grafts with the microscope; but Reverdin[*] has given the best description I can find. The sections were hardened in chromic acid, and colored with either carmine or ammonium picro-carminate. He saw the outer cells or cuticle separate in about 48 hours, and his sections, moreover, show that from the edge of the remaining graft-tissue a layer of epidermis[†] grows out over the ulcer, while another passes around the edge and *under* the graft. It is this last layer, he believes, which unites the graft to the ulcer by means of prolongations upwards into the graft and downwards into the ulcer. By the sixth day, this layer gives off large and small shoots, which penetrate the granulations, and have epidermic nests at their lower extremities (globes epidermiques). The layer which passes over the the surface of the sore terminates in a fan of smooth, spheroidal cells, with large nuclei. At the end of a week vessels connected with those of the ulcer appear in the graft, which, in turn disappearing, is replaced by tissue resembling granulation tissue. It has returned to an embryonic state. Reverdin believes,

  1st. That the graft adheres by a layer of epidermis which passes under it; and,

  2nd. That this epidermis acts by contact, exciting the embryonic surface to become epidermis.

6. *Nature of the cicatrix.*—It is a question whether, by this process, we have anything more than simple cicatricial tissue; for the sweat bulbs and hair follicles in the original graft disappear in the cicatrix. I have carefully looked for hairs, after taking pains to transplant them, but never succeeded in making them grow. To cover the whole of a large wound with cicatricial tissue is a

---

[*] In his essay cited above, Gaz. Hebdom., Jan. 1870.   [†] See wood-cut illustration.

great advantage ; for in such large ulcers the healing is very slow at best ; and often, as in the patient whose case illustrates this article, Nature is tired out, and ceases her work, leaving the patient uncured, and subjected, besides, to the constant drain from a large ulceration. Greater flexibility of the cicatrix is obtained if the grafting be close. Mr. John Woodman* claims that although true skin is *not* obtained (even the original graft disappears), the new tissue resembles it more than a common cicatrix does, and Bryant,† in his excellent article, says sensation appears earlier, and is more acute, in grafted tissue. I have never been able to verify the difference in sensation, but it is certain that the cicatrix is more flexible. In appearance, however, the new skin resembles that formed at the edges so nearly that it is impossible to draw the line of demarcation. The gradually increasing tension of a cicatrix, and its consequent feebleness, as it advances toward the centre of a large sore, is well known ; and when tension becomes sufficient to cut off the blood supply, cicatrization ceases. This is certainly largely prevented by grafting, and although such a cicatrix must be most carefully watched until it becomes strong, we may say that, being under less tension, it is probably better ; certainly quite as good as the one unaided Nature provides ; while the amount of time and vital force saved, in cases which would recover without it, must be immense.

7. *Action on surrounding tissue.*—One of the most marvellous effects of grafting is the very decided stimulation which the whole wound receives. Most marked near the grafts, it is seen even on distant parts of the ulcer. The wound on the hand of the patient whose case illustrates this article was decidedly improved by grafting on the breast. But its action is best watched by placing several grafts in a row along the edge of an ulcer, when each successful graft will be marked by a peninsula of skin jutting out toward the centre ; the *surface* as well as the edges of the wound are made brighter and more active.

8. *Method of growth.*—The origin of the cells by which a graft grows is doubtful. Reverdin claims its action to be catabiotic, because he has never seen its cells proliferating. The new epidermis would thus be made up of cells from the granulating surface which had been so changed in form by the presence of the graft, that they assume the appearance and function of epidermis. If proliferation takes place, the new cells are simply nourished by the granulations, grow, and reproduce themselves.

By means of this method large wounds are no longer hopeless affairs to the surgeon. Given a healthy surface, and abundance of material for grafting, and almost any surface can be covered—while the time and strength saved the patient is incalculable. Its application is not limited, however, to large surfaces ; but wherever it is desirable to prevent contraction or scars it may be used. An ulcer half an inch in diameter may be covered with grafts, and entirely closed in a day or two. Skin-grafting has not yet received the attention and general application which it really deserves.

* Monograph by John Woodman, published by Churchill, 1873.
† Bryant's Surgery, article Skin-grafting, page 130.

IX DUODENAL ULCER

(Case of Dr. F. H. Campbell)

CASE OF DUODENAL ULCER.

BY FRANCIS WAYLAND CAMPBELL, M.A., M.D., L.R.C.P. LOND.

Professor of the Theory and Practice of Medicine, Medical Faculty, University of Bishop's College, Montreal.

Duodenal ulcer is not a very common disease, although not so rare as some authors would lead us to infer. Within the past six years, several cases have been met with in Montreal, and this one—the history of which I will very briefly detail—was possessed of several interesting clinical facts. W. B., at the time of his death (1878), was 38 years of age, and being an intimate friend, was almost constantly under observation for over ten years. Just before forming his acquaintance he commanded a ship trading to the West Indies, which was wrecked, and he suffered much hardship. He then became a life insurance agent, and was most irregular in his meals. Flatulence about dinner time was often distressing, and whenever his feet were cold, a severe pain was located about the pyloric opening of the stomach. This disappeared when the circulation in the feet increased their temperature. In 1871, while absent from home, and shortly after an attack of flatulence and pain, he vomited at least a pint of blood. He, however, made a rapid recovery, and continued in fair health till the winter of 1874-75, when, on hurrying one evening to overtake the horse cars, he was seized with sudden faintness, accompanied by great blanching of the features. Some hours afterwards he passed a very large black stool, which was soon followed by a second. Several times during this winter, after more than usual exertion, he had fainting turns, followed by either vomiting of blood or its passage per rectum. I diagnosed gastric ulcer, and believed it was situated just inside the pylorus, which opinion was confirmed by several medical friends, who saw the case in consultation. Rectal alimentation was suggested, but was refused. Under a bland, but nourishing diet, he slowly regained fair health. As soon as he returned to work, whenever he became irregular in his meals, the epigastric pain would return, but it was not till the summer of 1876, that any vomiting of blood again occurred. With this exception he enjoyed apparently good health till the winter of 1877, when pain became again a prominent symptom, but without loss of blood. A milk diet rigidly carried out for a couple of months, and then for a couple of months more with the addition of either sago or arrow-root, not only gave ease but brought him into summer in such excellent condition, that I hoped the worse was passed. Unfortunate exposure, while heated, to the cool night air, brought on a chill, which was soon followed by intense gastric pain. The stomach became obstinate--nothing, not even the blandest food, could be tolerated in any quantity, although the appetite was ravenous. This condition lasted all summer. In the autumn of 1877, he had several large bloody stools, with a fixed burning pain, just below the point of the scapula. This latter being a new symptom, Dr. R. Palmer Howard, of McGill University, now saw the case with me, and, while agreeing as to the nature of the disease, was inclined to the opinion that the ulcer was duodenal. Rectal alimentation was insisted upon, as being necessary to sustain life, but the patient refused to submit. He left immed'ately for New York, and at my suggestion, took advice while there. The diagnosis given was gastric neuralgia. Strange to say, from the moment he obtained this opinion, he was able to eat everything that came before him, without pain, and so he continued for several days till he reached Boston, when again advice was sought, and again a diagnosis of gastric neuralgia given. On his return to Montreal, a few days later, this immunity from pain after eating continued, but it was soon broken, and the pain returned—worse, if possible, than before. Firm in the belief that I and my confreres were wrong, he placed himself under the care of a homeopathic practitioner, who diagnosed gastralgia. Repeated hemorrhages occurred, and the pain after even a small quantity of food was agonizing. I was often sent for during the night to relieve him by a hypodermic injection. On the 28th of January, 1878, he died from exhaustion, due to hemorrhage. The autopsy was performed by Dr. Osler, Pathologist of the Montreal General Hospital.

AUTOPSY.

Duodenum.—The part immediately outside the ring was much narrower than adjacent regions, measuring only 3.7 cm. About 10 m. from the pylorus there is an oval ulcer on the mucous membrane 2.5 by 1.8. cm., extending in direction of axis of gut, and occupying chiefly the posterior section of the tube. It is deep, with rounded edges, which, toward the upper and back part, are undermined for about 6 mm. In places the floor of the ulcer is quite 6 or 7 mm. below the level of the mucosa, and presents a tolerably smooth, fibrous appearance. The head of the pancreas forms the base of the lower three-fourths, the upper part is protected only by the thin muscular walls of the first piece of the duodenum, the peritoneal surface of which, at the site of the ulcer, is puckered and cicatricial. Immediately in the centre of the floor, is a small, dark, blood-stained elevation, consisting chiefly of fibrin. On injecting water through the hepatic artery, small clots are washed out at this point, and the water flows freely into the ulcer, through an opening in the gastro-epiploica dextra 2 mm. across and with smooth edges. The papilla of the bile duct is 6 cm. below the ulcer.

[58]

TWO CASES OF
FISTULA OF THE ANTERIOR PORTION OF THE URETHRA.

SUCCESSFUL TREATMENT BY PLASTIC OPERATIONS.

BY CHARLES McBURNEY, M.D.,

Visiting Surgeon to St. Luke's and Bellevue Hospitals; Instructor in Operative Surgery at the College of Physicians and Surgeons, New York City.

William Blank, forty-seven years of age, entered my service at St. Luke's Hospital in November, 1875. He gave a history of many claps, and stated that, several years before his admission to St.

Fig. 49.

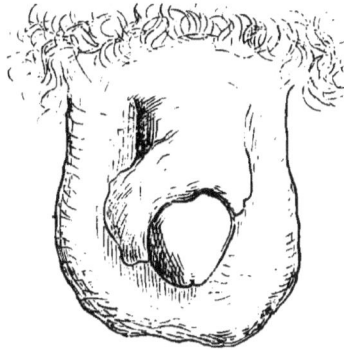

Fig. 50.

Luke's, a small opening, not larger than a pin's head, had appeared on the under surface of the penis, between the glans and the scrotum. Through this opening urine passed whenever the bladder was emptied. The patient also stated that the closure of this opening had been attempted by several surgeons, and that he had been etherized and operated upon no less than seven times. The exact nature of these operations I was unable to ascertain. Every operation had failed, and, apparently through suppurative action and the undermining of integument, the penis had been brought to the condition represented in Figs. 49 and 50.*

Fig. 49 represents the penis turned up against the abdomen. The fistula is marked by the letter A. It was three-quarters of an inch in length, and embraced the entire width of the urethra. With the exception of a small portion next to the scrotum, no integument remained on the under surface of the penis. On the left side (looking at the penis from below), was a large loose flap of integument, which hung loosely from the side of the penis, and on the opposite side one could just see the edge of the dorsal skin. The rest of this inferior surface was purely cicatricial. The loose flap of skin is marked B. Fig. 50 represents the dorsal aspect of the penis. Under these conditions it was evident that no attempt to close the fistula by *one* operation could succeed. I determined that I must first surround the opening with healthy integument before I could attempt to close it. This was accomplished on December 7th, with the assistance of Doctors Geo. A. Peters and Thomas T. Sabine, and in the presence of others. Having first opened the urethra in the perinaeum, in order to provide against the passage of urine over the seat of operation, I removed a thin layer from the entire cicatricial surface shown in Plate 49, including also the inner surface of the

Fig. 51.

loose flap of skin. The under surface of the penis was thus rendered completely raw, nearly back to the scrotum. The loose flap was easily brought to the edge of the fistula, and carefully stitched to that edge. I then cut a large square flap of skin from the left side and dorsum of the penis, and twisted it so as to bring it to the left edge of the fistula, where it was also stitched in place. These flaps were also stitched to one another above and below the fistula, and also to the base of the glans penis. The fistula was left, of course, as large as before, but was surrounded with healthy integument. The condition after this operation is shown in Fig. 51. C, the fistulous opening; B, the loose flap stitched

* For the original drawings of this case, and also for great care during its treatment, I am much indebted to Doctor Robert Abbé, who was then House Surgeon at St. Luke's Hospital.

to the right edge of the fistula, and A, the flap twisted from the left side and dorsum. The penis was then enveloped in lint soaked in carbolized water, *and subsequently the urine was regularly and carefully drawn by introducing the catheter through the perinæal wound.* At the end of the first week union was found to be complete, with the exception of the attachment on the right side, where a small portion of the flap broke away, and remained slightly undermined at the extreme edge. All granulating points having cicatrized, I took the second step in the case on January 17, 1876. The perinæal wound had already closed, and I again freely opened it. It was necessary to construct a floor to the urethra, which was accomplished in a manner similar to, and suggested by, the operation described by Julius Von Szymanowski, in his Handbuch der operativen Chirurgie, page 405. On the right of the fistula, where the skin was already slightly undermined, the knife was introduced so as to separate the skin still more, and form a pocket perhaps half an inch in depth. On the left of the fistula a large square was cut with its attached base at the edge of the fistula. This flap was dissected up, and turned over so as to present integument to the urethra, and was moreover tucked under the integument of the opposite side, its edge being carried to the bottom of the pocket already described, and there retained by means of sutures carried directly through the skin. The undermined edge of the right side was then drawn over the inverted flap (raw surfaces being thus brought in contact), and all adjacent edges were carefully stitched together. A floor was thus formed to the urethra, this floor being composed of *two thicknesses of skin.* I did not expect to obtain at this time a complete closure of the fistula, but only to cover the principal part of it. Union took place wherever raw surfaces had been brought in contact, but there still remained a small round opening about one-eighth of an inch in diameter at the base of the glans, and concealed by the overhanging edge of the new floor, and also a small linear fistula at the posterior edge of the same. The condition of the penis after this operation is shown in Fig. 52, B, indicating the position of the anterior opening (the opening itself being concealed by the overhanging skin), and A, indicating the posterior linear opening. Tempted by the small size of the anterior opening, I next made an attempt to close it by paring the edges and drawing them together with fine sutures. This was, as might have been expected, a complete failure, the stitches tearing out, and the opening being left rather larger than before. I had neglected to adopt the plan which is essential to success, namely, to supply new tissue to fill up the opening, and to

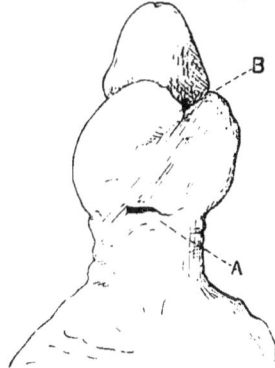

FIG. 52.

supply that new tissue in as great thickness as possible. This plan I carried out in the next attempt, made on April 14th. The perinæal wound had again closed, and was now again freely opened. There was, of course, no foreskin, but the integument posterior to the opening overhung it to such an extent, that I could turn it back, and mark out upon the surface presented to the fistula an area about one-third of an inch square. This surface was rendered raw, with the exception of the central portion, which was intended to fill the opening, and thus present integument to the urethra. A corresponding raw surface was made anterior to, and around the fistula, so that on again turning the skin forward raw surfaces were brought in contact, while the central unrawed portion filled the opening. Moreover, the anterior end of the central unrawed little flap was partly dissected up and tucked within the urethra beneath the anterior edge of the fistula, this portion of the urethra having been previously carefully scraped with the knife. The cuts Nos. 53, 54 and 55 show the different steps of this operation. Fig. 55, A, the overhanging edge of skin turned back from over the fistula; D, the raw surface made on

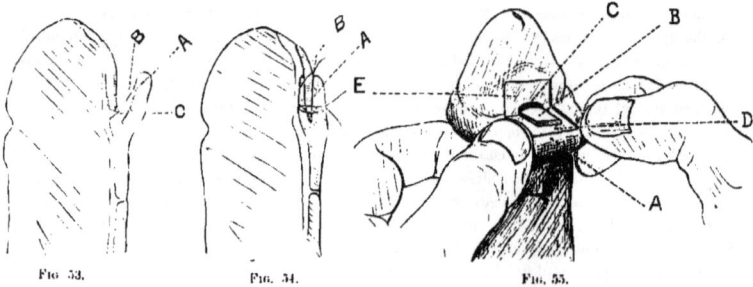

Fig. 53.   Fig. 54.   Fig. 55.

this flap; B, the central unrawed portion; C, the fistula; E, the raw surface made anterior to the fistula. Fig. 53, sectional view; C, the fistula; A, large skin flap; B, central unrawed portion. In Fig. 54, the flaps are seen in position, B indicating the small flap of skin tucked into the urethra. This inner flap was held in position in this manner. Two fine sutures, each armed at both ends with curved needles, were first passed through the end of the flap, the needles were then passed into the urethra and out through its floor anterior to the fistula. These sutures being then drawn tightly and knotted, the flap was drawn into place, and firmly held there. The adjacent edges of the outer raw surfaces were also firmly stitched. Union took place rapidly and completely, and there remained only the posterior linear opening. The situation of this is indicated in Fig. 52, A. This fistula was also closed at the first attempt, on June 29, 1876. The perinæal wound had healed and was again opened. I first marked out a small flap on the right, having its base at the edge of the opening. The rest of the margin of the fistula was rendered completely raw for the reception of another flap, to be described. The small flap being then turned over, integument was presented to the fistulous opening, and the edges of the flap having been tucked within those of the fistula, it was stitched in place. To cover the raw surface which surrounded the inverted flap, as well as to cover and help nourish the inverted flap itself, a long flap was cut from the scrotum, which was easily drawn forward and stitched in place. Before the stitches were applied to the edges of the scrotal flap, several fine sutures were passed through its cutaneous surface from without inwards, then through the raw surface to which it was applied, and then out again to the cutaneous surface. These sutures being then drawn tightly, and knotted, served to hold the raw surfaces firmly together. Lastly, stitches were applied all about the edges of the scrotal flap. A piece of tape was passed beneath the scrotal flap posterior to the fistula, in order that the flap might at a later period be more readily cut free from the scrotum. The various

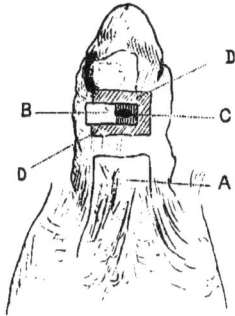

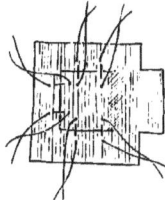

FIG. 58.

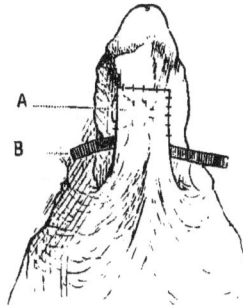

FIG. 56.

FIG. 57.

steps of this operation, as well as the method of applying sutures, are clearly indicated in Figures 56, 57 and 58. B, the small flap which was turned over to cover the fistula; C, the fistula; D, the surface around the fistula which was rendered raw; A, the scrotal flap.

The closure of this fistula also took place readily, and when cicatrization was complete, I divided the band of scrotal integument at the point where the tape had been passed. The perinaeal wound

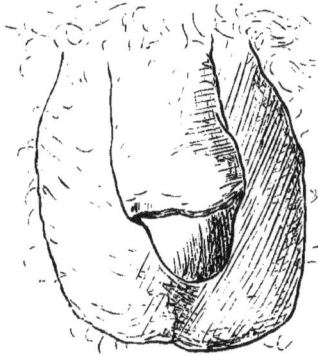

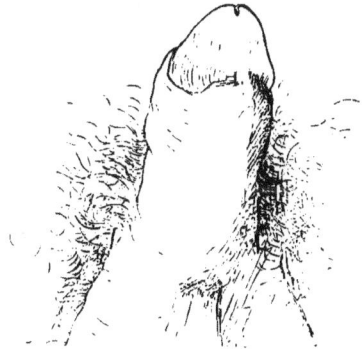

FIG. 59.

FIG. 60.

healed in less than three weeks. This patient complained for some time of a downward curvature of the penis during erection, but this defect rapidly disappeared without treatment. (Figs. 59 and 60.)

On May 18. 1877, about one year after the last operation, I made a careful examination of the case, using Otis' Urethra meter. The penis was found to be perfectly straight, and the patient stated that during erection no curvature occurred. In the urethra some differences in calibre were found. At one and one-half inches behind the meatus, the urethra meter showed a calibre of 25 F; at two inches, 27 ; at two and a half inches, 30; at three inches, 31 ; at four inches, 28. No. 26 steel sound passed easily into the bladder without causing discomfort.

I saw this patient last in June, 1881, and found that he was not conscious of any symptom referable to stricture. He stated that the penis, when erect, was somewhat shorter than it had formerly been.

CASE II.—Geo. W. Colby, forty-eight years of age, a planter from Georgia, was admitted to St. Luke's Hospital on April 7th, 1879. Five months before admission, a phagedenic sore of venereal origin appeared at the base of the glans penis, just at the right of the fraenum. At the end of two months this sore had opened into the urethra. The ulceration slowly healed after this, leaving an opening just at the right of the fraenum, the fraenum being nearly destroyed. The fistula was surrounded by a mass of dense cicatricial tissue. This cicatricial tissue caused a decided bending of the glans downwards and to the right, when erection took place. The meatus was contracted to No. 20 F, and almost all of the urine passed through the fistula. On the 8th of May I enlarged the meatus, and on the 27th of the same month I divided strictures which existed at 2¼, 2½ and 3½ inches behind the meatus until steel sound No. 34 F. could be passed readily into the bladder. The patient left the hospital for a while, but returned for the cure of the fistula.

The condition of the penis at this time was as follows: Just at the base of the glans at the right of the median line was a fistula opening directly into the urethra, and large enough to admit a No. 20 French bulbous bougie. The opening was surrounded by dense cicatricial tissue. Nothing was left of the fraenum but a thin bridge of skin. The foreskin was fortunately very abundant, and the urethra had been rendered perfectly patent by the division of the strictures.

August 22d I operated upon the fistula. The urethra was first opened in the perinaeum to provide against the passage of urine over the seat of operation. The foreskin was then pulled tightly backwards, and the surface around the fistula to the extent of one-half inch square was completely rawed. This surface of course included the base of the glans. Posterior to this, on what was the inner surface of the foreskin, a corresponding area *with the exception of the central portion*, was also rendered raw by dissection with the knife. The central portion was left undissected for two reasons. In the first place I wished to present integument to the urethra, and so avoid cicatricial contraction within the urethra; and in the second place I intended to push the end of this central portion within the urethra under the anterior edge of the fistula. The parts were then arranged as I have attempted to indicate in Figs. 61 and 62. The foreskin being now simply rolled forward, the lines A'C' and B'D' were doubled on themselves at their centres, so that the edges C'D' and A'B' were brought in contact. The central unrawed portion not only covered the fistula, but its extremity, which was now anterior, was somewhat detached, and was tucked under the anterior edge of the fistula, which edge had previously been scraped with a small knife. This small portion was held in place by two fine sutures carried through the floor of the urethra and then through the outer flap. All adjacent edges were then carefully stitched together with the finest silk. The penis was enveloped in lint soaked in carbolized oil. After this the urine was carefully drawn through the perinæal opening several times daily.

Four days later a fresh dressing was applied.

August 29th, several stitches were removed, and on September 5th, the first time that a thorough examination was made, union was found to be complete throughout.

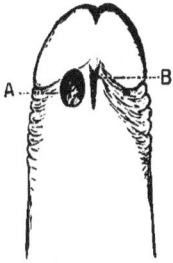

Fig. 61.

September 12th. All of the urine passed naturally through the penis.

September 18th. The perinæal wound had entirely healed. On discharge from the hospital this patient still noticed a slight bending of the glans downwards and to the right when erection took place. This was due to the extensive cicatrix left after the healing of the original ulceration.*

It is believed by the writer that the success attending the operations described was due to several causes. In the first place, new tissue was supplied, in the form of a freely movable

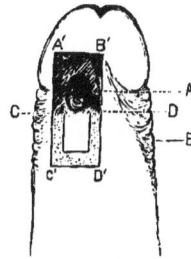

Fig. 62.

flap to close each fistula operated upon. In one instance this was not done, but the attempt was made to simply stitch the prepared edges of the fistula together, and the operation completely failed.

In the second place two thicknesses of skin were used (integumental surface being always presented to the urethra), so that greater vitality was secured to the inverted flaps.

Thirdly, the urethra was always opened posterior to the situation of the fistula, thus completely diverting the course of the urine during the process of healing.

It is well known that while perinæal fistulæ, due to disease, sometimes heal with great difficulty, those produced with the knife through healthy tissue heal readily. With regard to the use of sutures, it may be said that the very finest were always used, and were so applied as to obtain the most complete apposition of surfaces to one another.†

* For great care shown in the after treatment of this case I desire to express my indebtedness to Doctor John F. Ridlon, at that time House Surgeon at St. Luke's Hospital.

† Read before the New York Surgical Society, November 22d, 1881.

TWO CASES OF GUMMOUS IRITIS.

BY F. R. STURGIS, M.D.,

Ex-Professor of Venereal Diseases in the Medical Department of the University of the City of New York, &c., &c.

———

These two histories are given on account of the rarity of the affection and the important consequences which it entails.

CASE I.—S. B., æt. 30, entered Charity Hospital August 30, 1874. Initial lesion occurred in 1871, followed by papular, papulo-squamous and pustular syphilides of the skin, alopecia capitis and osteocopic pains. She had two miscarriages during the early months of pregnancy.

At time of entrance she had a superficial ulcerating syphilide of the anterior nares and upper lip, and shortly after, September 5, she developed a double kerato-iritis and cyclitis, with violent photophobia, lachrymation and supra-orbital pain. V. reduced to perception of light. Under active treatment all symptoms improved until October 5, when the irido-cyclitis returned in both eyes, and within 48 hours a growth was perceived [Pl. xi. Fig. 66] in the right eye, apparently springing from the uvea iridis, occluding the pupil. V. R. E.=0. V. L. E.=perception of light. Associated with this was an intense hemicrania. No ophthalmoscopic examination could be then made. Mercurial inunction and large doses of the iodide of potassium caused the disappearance of the gumma. An ophthalmoscopic examination showed an atrophic stage of choroiditis in both eyes. V. not taken from this time from oversight. Later on she had several relapses of irido-cyclitis, but no recurrence of gummata. She subsequently left the hospital much improved, but not well of her syphilis.

(Note. She died two years afterwards in the Hospital, of syphilitic cachexia, but had had no return of the gumma iridis.)

CASE II.—M. C., æt. 35, entered Charity Hospital September 30, 1875. Initial lesion appeared September, 1873, followed by papular and pustulo-crustaceous syphilides and alopecia. At entrance she showed a pustulo-crustaceous syphilide, one of the ulcerations being seated at the outer canthus of the left eye. This organ also showed a well-marked iritis, together with remains of a conjunctivitis, attended with intense lachrymation, photophobia and pain in the supra-orbital and temporal regions. Pupil L. E. was occluded; V.=0; V. R. E.=1. On dilatation the occlusion was found to be due to a gumma [Pl. xi. Fig. 67] which sprang by a broad base from the uvea iridis. K. I. internally and Ung. Hg. externally, caused the disappearance of the growth in three weeks' time. An ophthalmoscopic examination showed no choroiditis nor retinitis.

X. CONGENITAL KERATOMA

CASE OF DR. GEORGE G. WHITLOCK

A CASE OF DIFFUSE CONGENITAL KERATOMA

(CONGENITAL ICHTHYOSIS).

BY GEO. G. WHEELOCK, M.D.,

Attending Physician to St. Luke's and the Nursery and Child's Hospitals, New York City.

The female child, of whom the photograph on the opposite page is a very faithful representation, was born at the Nursery and Child's Hospital, in New York City, May 20, 1882. The mother, a primipara, was a young and healthy woman, and, as far as could be learned, the father was a healthy, robust man, free from any constitutional disorder or skin disease. The mother certainly bore no trace of any constitutional taint, denied any venereal history, and showed no sign of present or past affection of the skin. No history of fright nor any unusual occurrence during pregnancy, no abnormal pains nor sensations could be gathered from the patient. The first stage of labor lasted about thirty (30) hours, during the last three of which no progress was made after the membranes had been ruptured by the house physician which was about two hours before I was called to see the patient. Upon examination, the os was found about two-thirds dilated, but thick and œdematous, and although it was evidently a vertex presentation, the position could not be made out, for the presenting part was crossed and broken up by various sulci running in every direction, with rough, irregular edges, which gave to the finger nearly the impression of carious bone, the smooth, hairy scalp being nowhere perceptible.

The forceps were applied through the os, and the child was with considerable difficulty extracted, there seeming to be a total lack of lubrication, and consequently great friction between the fœtal and maternal parts. As the head was born, a thick plate of skin two inches square was detached and escaped with the head, and it was then supposed to be a firm layer of vernix caseosa. At first the child had the appearance of a dead fœtus with macerated epidermis, and I was greatly astonished to see it shortly begin to breathe, and before long to cry feebly. Its appearance was horrible in the extreme. It was covered from head to foot with a skin like leather, which was deeply fissured in all directions, and broken up into plates like an alligator or an armadillo. Along the neck and abdomen, many of the plates were completely separated, so that the true skin of a bright, strawberry color showed through, while the dingy yellow plates of epidermis separated and rose and fell with each respiratory act. The child was apparently of normal muscular development, and had reached full term. There was nothing wrong about the placenta, which came away readily, and bore no marks of fatty or calcareous degeneration. After birth this appearance increased, as the skin dried it became of a bright, chrome yellow, and the plates were more and more detached by the motions of the child. It lived only six hours.

This condition has been described under various names by different authors. In 1843 Dr. (after wards Sir James) Simpson contributed an article to the Edinburgh Medical Journal, on "Congenital Ichthyosis," describing a case he had seen, which description corresponds almost exactly with the present case. The most complete and exhaustive description may be found in Stricker's "Medizinische Jahrbücher" for 1880, by Dr. Kyber of the Russian Navy. He calls it a "Diffuse Congenital Keratoma," which name would seem to be the proper one, as the abnormity consists chiefly in an enormous hypertrophy of the corneal layer of the epidermis, with excessive development of the papillæ of the corium. The epidermis has the appearance of rather coarse kid, of a yellow color, and of almost the firmn ess and consistency of cartilage. Where it is preserved intact, it is smooth, without any prominences or spines such as are described in many cases of ichthyosis, and has no scales such as appear in some forms of that disease. To judge from its appearance, it is purely the mechanical result of the growth and development of a child within an unyielding case of dense skin, as separations and fissures have occurred in all directions, and are most numerous where flexion and extension are performed during intra-uterine life, notably on the neck, anterior part of abdomen, and at the joints of the extremities. These fissures leave islands of dense skin of varying shapes, bounded by shelving margins which have a shiny fibillated appearance, and meet the similar margins from adjacent islands, and at their line of junction is found the thinnest portion of the covering. These plates are largest on the back and buttocks; smallest on anterior part of the trunk and the extremities. The scalp, with here and there some hair about its greatest circumference, is extensively fissured over the vertex and occiput; the ears are incompletely formed, very hard and firm, and bound down flat to the head, while no meatus externus can be found. The nose is completely flattened, and shows only the prominence of the alar cartilages and two nasal apertures. Over the nose the skin is intact, but fissures exist on either side, and extend round the mouth, which is large, held widely open, is elliptical in shape, with everted lips and a large, protruding tongue. The eyebrows are wanting; there is ectropion of the upper lid of the right eye, showing a bright red cushion of mucous membrane filling the aperture of orbit; the left upper lid is not everted, and the eyes underneath are apparently normal. On the neck, especially behind, the plates are almost completely detached, uncovering the true skin. Circular fissures extend around the wrists running deeply into the underlying tissues, but not uncovering the carpal bones as has been observed in some similar cases. The hands are very much deformed, resembling the paws of an animal; the epidermis extends only to the first phalanx, beyond which the fingers are clubbed and armed with rudimentary nails resembling claws. The feet are almost cylindrical in shape—hard and swollen, completely covered with epidermis, without fissures, and bearing each five rudimentary toes. The external genital organs are perfect, with vulva and hymen of normal appearance and meatus urinarius. The anus is perforate, though apparently rather small. The autopsy revealed the presence of fluid blood and fibrine in the abdominal cavity, with ecchymotic spots on the intestines. The kidneys on section intensely congested, but otherwise normal. Spleen congested, and liver dripping with blood. Lungs in state of partial atelectasis. Ecchymotic spots on the pericardial surface of heart. Brain the seat of many small hemorrhages with one quite large clot on the surface of posterior part of left lobe.

This case will be more fully and accurately described with microscopical pathological changes in the October number of the American Journal of Obstetrics.

Fig. 65

Fig. 66

GUMMOUS IRITIS

(Case of Dr. F.R. Sturgis)

XI PAPILLOMA OF PHARYNX

(Case of Dr. J.O.Roe)

A LARGE RECURRENT PAPILLOMA OF THE PHARYNX.

BY J. O. ROE, M.D.,

ROCHESTER, N. Y.,

Fellow of the American Laryngological Association; Member of the American Medical Association; of the Medical Society of the State of New York, &c.

Mrs. Thomas Tousey, æt. 61, was referred to me by Dr. Carpenter of Pittsford, February 14th, 1882, with the following history. Her throat had been very sensitive for many years. One year ago she noticed a sensation of dryness in the throat, followed six months later by slight dysphagia, which progressed until she was unable to take anything but liquids, and these only with great difficulty.

Four months before consulting me her voice began to have a muffled nasal sound, which, when I saw her, was marked. Respiration was unaffected.

Her general health has been poor for ten or fifteen years past, and recently she has failed rapidly, owing to her inability to take sufficient food.

About six weeks before I saw her, Dr. Carpenter observed a small growth in the lower part of the pharynx, which increased rapidly in size; indeed, at the time I saw her, the tumor had grown so large as almost to fill the lower pharyngeal space.

It involved the entire posterior wall of the pharynx from opposite the lower border of the soft palate (slightly higher on the left than the right side), to nearly the upper orifice of the œsophagus, about opposite the upper border of the cricoid cartilage. It involved both posterior pillars of the fauces, extending to their anterior border, but not any portion of the larynx. Its right side was a little thicker and more projecting than the left.

It had a roughened, slightly lobular appearance, its meshes being filled with a slimy, tenaccous exudation (vide Plate XI.).

On account of the rapid growth of the tumor and the dysphagia I advised its immediate removal, which was done February 16, 1882, with the galvano-cautery.

I had hoped to encircle the growth with the galvano-cautery wire and remove it entire, but was unable to do so, on account of the breadth of the base.

The knife used was constructed with a curved stem, the blade turning laterally so that it could be swept around behind the tumor, peeling it off, as it were, from the pharynx, but, owing to the friability of the tumor, it came away in fragments.

No hemorrhage followed, and but little inflammation or pain, although no anæsthetic was given. Immediate improvement in swallowing followed, and solids could be taken quite readily.

March 7th. The tumor has been growing again rapidly, and has attained to nearly half its original size. Owing to the absence of hemorrhage I now removed the entire tumor with a pair of curved scissors, and cauterized the surface with fuming nitric acid, and later on with chromic acid, which was applied about once a week for a month, with the effect of keeping the growth in check, but not destroying its vitality.

I then decided to follow up this treatment more vigorously. Accordingly I applied the chromic acid every second day to all parts of the growth, scraping off the slough each time before making a fresh application, and removing with the scissors any portion which resisted the action of the acid, or seemed to grow in spite of it.

After five cauterizations the patient's throat became so sore as to necessitate an intermission of treatment for ten days.

At the end of this time inspection of the throat showed what appeared to be an outgrowth of the tumor, but it proved to be only a slough, which I removed, leaving a clean surface beneath.

The only remaining portion of the growth was a small nodule on the inner surface of the right posterior pillar above, and at the lower portion of the pharynx. These I cauterized with the chromic acid, and they have disappeared, leaving the pharynx free from any evidence of re-appearance of the growth.

The diagnosis which was made prior to the removal of the growth, viz. : true papilloma, was fully confirmed by the subsequent microscopic examinations, the details of which it is unnecessary to state, as the character of these growths is so well known.

Remarks.—The behavior of this growth in regard to hemorrhage is quite exceptional to that of ordinary papillomata, as, owing to their vascularity, free bleeding is the rule on cutting or scarifying them. It was for this reason, and from the extent of surface which it covered, that I used the galvano-cautery at the first operation, so as to avoid depleting the patient, which it was feared direct excision without the cautery would be liable to do.

In this connection I would call attention to the superiority of chromic acid over all other caustic preparations for the destruction of indurated tissue or of the base of tumors after abscission. It causes but little pain, even when applied to a raw surface, while the application of other caustics to denuded surfaces is exceedingly painful, and is followed by considerable inflammation. It seems to have a drying and shriveling effect upon the tissue to which it is applied, and at the same time to exercise a healthful alterative action on the tissues beneath.

Note.—Since the above was written the growth has slowly reappeared at the lower portion of the pharynx and entrance of the œsophagus, attended with increased dysphagia.

THE THERAPEUTIC USES OF RUBBER TUBING.

BY WM. M. CHAMBERLAIN, M.D.,

Visiting Physician to Charity Hospital, Blackwell's Island, New York.

The therapeutic uses of rubber tubing, together with nearly all the appliances figured or described in the following paper, was presented by the present writer to the State Medical Society, in 1874, and to the New York Journal Association at the meeting of March 13th, in the same year.

As the paper itself was not offered for publication, and the reports in the journals* did not show drawings of the apparatus, the methods there proposed did not obtain a general currency.

Several individuals have reported to the writer that they have employed the system, and various devices in the same line have from time to time been since presented.

Within the last year, however, the same ideas have been reproduced in Vienna, and a cumbrous and extensive system of apparatus, protected by an international patent under the name of "Leiter's Tubes," was presented to the International Congress at London, as a novelty.

"Leiter's Tubes" differ from those proposed and exhibited here in 1874 only in the material: they are made of lead, instead of rubber.

The general attention which they have received in Europe, the fact that it was proposed to obtain a patent for them here, and the opinion that our apparatus is much better than the foreign, led the writer to present the matter anew to the Academy of Medicine, on April 20th; and to comply with the request of this journal to furnish an illustrated paper upon the subject.

The "Therapeutic Uses" considered were irrigation in its various applications, and the control of inflammation and general pyrexia by withdrawing or adding heat by currents of cold or warm water conducted in tubes over the surface.

The superiority claimed for the American method lies in the following particulars:

1. Any required appliance can be *extemporized* and accurately adapted to the precise indication, which is impossible in the fixed metallic system.

* Transactions of N. Y. St. Medical Society for 1875, p. 16; Medical Record, Vol. IX. p. 96; Medical and Surgical Reporter, April 11th, 1874, p. 331.

[71]

2. The rubber tubes are very much lighter and more comfortable, by reason of their lightness, and of their less rapid conduction.

3. They are less liable to accident, since a fall or any undue compression will destroy the metal tubes, or their efficiency.

3. They are not patented, are less expensive, and as they may be prepared when and as wanted, do not involve the purchase and storage of many cumbrous pieces.

The primary irrigating apparatus is shown in Fig. 71. A tube of white vulcanized rubber, five feet long, three-sixteenths of an inch calibre. One end is slipped over the stem of a tubulated weight (I have found the forms 1, 2 and 3 in the shops), and the other end is fitted with a small brass stop-cock (4). The tube is to be filled, using the cup-like weight as a funnel, or by suction, the stop-cock closed, and the weighted end dropped into a bottle, pitcher, pail or tub, according as little or much water is needed. Upon opening the stop-cock the water will flow with a force proportioned to the difference in level between the two ends of the tube.

For some uses a bottle with tubulars at bottom (5) is convenient; the tube will then be always filled by gravity.

Posterior Nares.—Irrigation of the posterior nares (Weber's Douche) is less in vogue now than it was ten years ago, on account of the accidents which have followed the entrance of water into the middle ear. Nevertheless, the method is retained by some of the most eminent throat specialists, and the danger is held to be infinitesimal if the patient has been sufficiently educated to keep the head bowed forward, the mouth wide open, and to control the impulse to swallow while the current of warm salt water is flowing with moderate force.

Fig. 71.

Stomach.—Irrigation of the stomach is rendered very easy and effective by the use of the long tube mentioned by Dr. S. O. Vanderpoel, and shown in Fig. 7↓, made of soft rubber with velvet surface, open, fenestrated at the distal end, and expanded into a funnel at the proximal. It passes easily through the œsophagus, aided by the swallowing effort of the patient. The tube is five feet long. When raised above the head of the patient, any given quantity (a pint) of water is poured into the funnel, it is then lowered, and the fluid returns at once, and the process is repeated as long as desired. Patients suffering from dilatation of the stomach, and requiring this process after each meal, learn to serve themselves in this way with comfort and facility.

Bladder.—In the same way the bladder is washed for cystitis. The soft catheter is passed, and connected with the stop-cock of the syphon tube. The bottle is held in the hand of the operator, the amount of force with which the current shall enter is graduated with utmost nicety by raising or lowering the bottle, and the sinking level in the bottle shows at each instant how much distension there may be. The return is rapid or slow, according to the distance of the bottle below the level of the bladder.

When the bladder has been emptied, the stop-cock is closed, the bottle filled with clean fluid, and the process repeated at discretion.

I do not know any circumstances under which I should employ the old and painful process of pumping water into the bladder. There is no need of a double catheter, or of one with two mouths. In fact, to fill the bladder, no catheter need be used in the male. A short glass tube, open at the end, may be passed through the fossa-navicularis, and the glans compressed upon it, by tightening the prepuce at the fraenum. With a pressure of five feet the current of water will generally open the whole tract of the urethra. Care

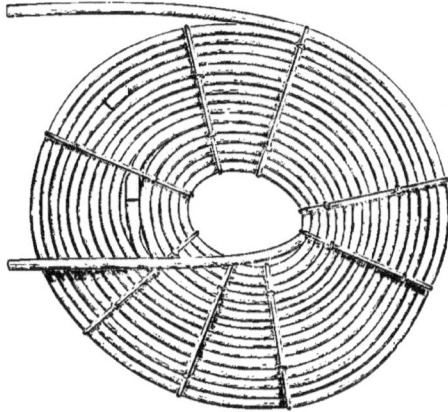

Fig. 72.

should, however, be used not to make too much compression of the glans. In two cases I have known an abscess of the glans to follow neglect of this precaution.

Vagina and Uterus.—Irrigation of the vagina, when managed by the patient, is commonly done with the fountain syringe. The India-rubber bag which is part of it is expensive, perishable and unclean. The weighted tube will better serve every purpose. When, as in puerperal cases

Fig. 73.

and in septic fever, this office must be rendered by another, the uterine tube of glass, eighteen inches in length, and half an inch in diameter (see Fig. 73) will be found very convenient. The patient need not be touched by the operator, who may stand erect at the bedside, pass the beak of the tube between the labia, and, with the gentlest pressure, to the os uteri. A little manipulation will now cause the beak to engage in the os, the outer end is then depressed and the beak slides on to the fundus, the current is liberated, and, having washed out the womb, returns to the bedpan placed beneath the patient. The value of the process in suitable cases cannot be over stated.

Bowels.—Irrigation of the bowels is greatly aided by the posture of the patient. Sims' position—semi-prone on the left side—still better, the "knee-chest" (Campbell), permits the reception of a large quantity (three to seven pints).

Professor Mosler of Greifswalde employs cold water thus to reduce the volume of the spleen, and to abate general pyrexia. The hydrostatic pressure thus obtained has for me twice reduced invagination of the intestine, the tumor having in each case advanced into the descending colon.

Surgical Operations.—Some surgeons like to have a tube, closed by a stop cock, hanging over the operating-table. A slender stream of carbolized water washes away the blood, clears the field of vision, and saves much sponging. The water runs in a gutter of the rubber table-cover to a pail beneath.

Ear Douche.—The very neat and effective ear douche which bears the name of Dr. Roosa deserves to be mentioned here.

II. Appliances for controlling local inflammation and general pyrexia.

The scope of this paper does not include a discussion of the value of cold and heat as therapeutic agents. They are the prime forces of nature, which represent Rest and Motion: Death and Life. "Ubi vis, ibi virtus." In the literature of the subject the names of Langenbeck, Traube, Wunderlich, Liebermeister, Ziemssen, Currie, Fox, and Murchison, are prominent. The most complete monograph is perhaps that of Esmarch upon "Cold in Surgery," translated in 1861, by the Sydenham Society. The review of the cases there published certainly seems to warrant the language in which the author delivers his opinion. "Of all the remedies which are at our command for the treatment of inflammatory processes, it is the most important; and without it I would rather not be a surgeon." Esmarch's appliances were rubber bags filled with iced water, or metallic boxes. Leiter's, as before mentioned, are leaden tubes. The objections to either are very considerable. To be effective, the bags must be continually changed, the degree of cold is constantly varying. Water will not lie upon the slope or crown of a convex surface; the bags are very heavy; in the constant handling, it is difficult to prevent leakage. The value of hot or cold applications, for the most part, depends upon their equable and continuous action. Nothing more easily defeats the end proposed than *alternate heat and cold*, expansion and contraction, chilling and flushing, which exhausts the elasticity of the vascular and aggravates the irritability of the nervous system. For the apparatus of rubber-tubing the advantages have already been enumerated.

The simplest form of the apparatus is shown in Pl. 71, where a current of water is seen running through a tube loosely wound around the forearm and wrist. This is applicable only to the extremities, and even then is not to be preferred. These parts may, and other parts must, be treated by means of tube confined in coils of various size and shape. Pl. 72. Given the tubing, it requires but little time to make such as may be required. Sitting before a table, covered with a woolen cloth, to prevent slipping, the coils are laid in circles, ellipses or parallelograms, as a sailor lays rope on the deck. Tubing is sold in lengths of twelve feet. As many as may be desired are joined by causing the opposed ends to meet over a bit of glass tube having a diameter larger than the lumen of the rubber tube. When enough has been laid, the border of the coil is drawn toward the edge of the table, and, while one section is confined by a *band* passing from the outer to the inner border, the rest of the coil is kept flat and in place by a ruler or a hook laid over it. The best way of making the bands, in the case of coils of large tubing, is to take smaller tubing cut in lengths equal to the diameter of the coil; slide into these lengths pieces of copper wire; bend the tube and wire like a hairpin, and pass one arm above and one below the surface of the coil. Tie these arms together at the inner border of the coil, and also between the strands of the coil; see Pl. 74. For coils made of smaller tubing, take the wire ribbon ("wire taste") used by milliners; bend like a hairpin over and under the coil, and sew between the strands.

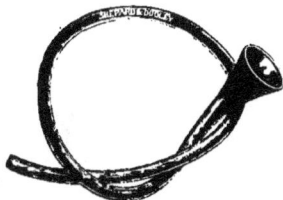

Fig. 74.

Pl. 75 shows a coil laid upon the head; one held over the eye, and one fastened around the throat. In the eye coil, the circles are maintained by a silver wire drawn through the

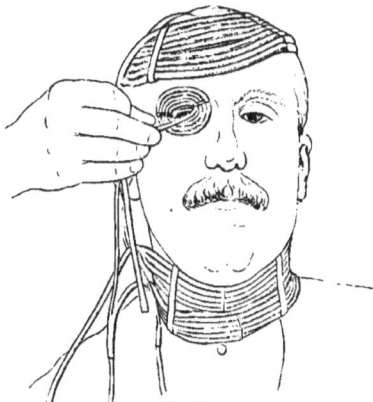

Fig. 75.

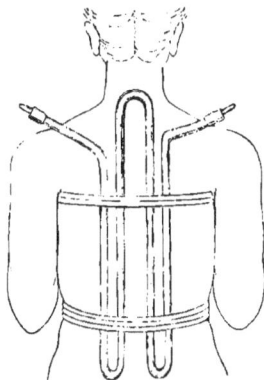

Fig. 76.

lumen of the rubber tube. Made thus,—*i. e.*, in either of the ways named,—the coil may be shaped to fit any surface, however irregular. The bent bands will retain it in any form, and it may be drawn out or compressed like a coil of wire.

Pl. 76 shows an appliance for the spine. For this purpose about ten feet of stout tube, one-half inch in calibre, should be used, and a brass wire, stiff enough to neutralize the elasticity of the whole, should be passed through it. It can then be bent into the shape shown in the figure; and the wire will not only retain it in such shape, but will prevent the tube from being closed by the weight of the patient lying upon it. The free ends of the tube may be fitted with india-rubber corks, perforated to pass a bit of glass tubing for connecting with the syphon.

The power of heat to reduce tension and relax spasms is well known. The warm coil then becomes serviceable in colic, ileus, pleurodynia, torticollis, lumbago, myalgia, arthralgia, &c., &c.; and in the more general conditions of collapse, the algid forms of cholera, and other diseases.

The warm coil will also be an excellent cover for any poultice, stupe, or fomentation, since it maintains both the heat and the moisture which are sought, makes a given temperature continuous, and saves in part both the labor and the risk of frequent change.

But it is for the use of *cold as a medicine* that the apparatus is most valuable.

The power of cold over zymotic processes is indicated by the ice treatment of diphtheritic and gonorrheal conjunctivitis, or the ophthalmia of the new-born. Upon what does the admitted success of that treatment depend? Does the cold assist the development of the microphytes or microzymes which destroy the epithelium faster than it can be reproduced? or is it simply the constriction of the capillaries by cold and its sedative influence on the nerve-loops of the conjunctiva? The question is interesting. The surgeons of the Woman's Hospital employ the cold coil upon the abdomen after laparotomy or ovariotomy to forefend and to treat those forms of inflamation, which, if they do not begin in sepsis, quickly end in it. Dr. Emmet reports a case in which the temperature was reduced from 104° to 99°, and kept down by the coil.

I have not found the coil equally serviceable in typhoid fever, although I have repeatedly seen the temperature, taken under the tongue, fall two or three degrees after circulating water at 70° through a coil laid upon the abdomen.

It cannot be expected that as a means of controlling general pyrexia this method will equal those of immersion, affusion (Kibbe) and the cold pack, but doubtless it may be made very valuable in cases where these methods cannot be employed, and in particular, in maintaining a reduction of temperature which has been secured by other means.

It is in local diseases that the most general and striking utility of refrigeration is seen. It is a question whether the inflammatory process can go on in a tissue cooled below a certain point; however this may be, it is certain that it is greatly restrained by cold. My own experience with it has been chiefly in injuries to joints—sprains, contusions, luxations: inflammation of deeper fibrous tissues, &c. In a recent case of penetrating and lacerated wound of the knee-joint—the wound having been closed antiseptically and the limb kept at rest on a splint, the cold was applied with a prompt relief of pain, abortion of all inflammatory symptoms, and when, after four days, the cold was removed, the joint was in apparently normal state and the external wound nearly healed. Many such cases may be found in Esmarch's monograph. He particularly commends it in disease of the vertebræ, and in spinal meningitis, as does Allibut in rheumatic cerebral meningitis, Niemeyer in pleurisy. In the tetanus which is so common on the eastern end of Long Island, the prolonged application of ice to the spine has been said to be followed by very great success.

Fig 74

Fig 75

Fig 76

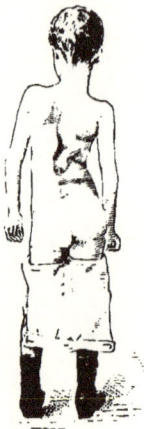

Fig 77

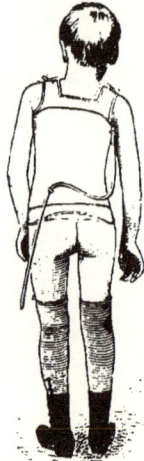

Fig 78

Fig 79

XII.—ELASTIC TENSION IN THE TREATMENT OF POTT'S DISEASE

(Case of Dr. M. J. Roberts)

ELASTIC TENSION IN POTT'S DISEASE.

EMPLOYED AS A THERAPEUTIC ADJUNCT IN THE MECHANICAL TREATMENT OF THIS BY SUSPENSION AND THE PLASTER OF PARIS JACKET.

By M. JOSIAH ROBERTS, M.D., New York.

Elastic tension is an active power which can be exerted in any direction and developed by a great variety of substances. I have elsewhere* considered the advantages of *elastic tension and articular motion in the treatment of chronic inflammations of joints*. I wish now to demonstrate the therapeutic value of *elastic tension* in rectifying trunkal distortions resulting from muscular weakness arising in the course of Pott's disease. I have used it in certain cases as an adjuvant to the now widely adopted plan of treatment introduced by Prof. Lewis A. Sayre.

All who have treated many cases of angular curvature, must have met with some in which, after the application of the plaster jacket or other spinal support, there was still present an undue protrusion of the belly forward, as in Fig. 74, or a lateral tilt, as in Fig. 78. Such distortions are, for the most part, due to debility of the muscles of the lumbar and the lower half of the dorsal regions. The improved position gained by suspension, or extension of the spine with the patient in the horizontal decubitus, is not maintained, no matter with how much skill the spinal support is applied. And for this reason, the power of the spinal lever is not applied below, or external to the weakened muscles. Were it practicable to carry a rigid spinal prop below the hip-joints down upon the thighs, the improved position of the spine would be readily maintained. This, however, would necessitate the immobilization of the hip-joints, and the crippling of the patient to such an extent as to absolutely prevent locomotion. In a case such as is represented in Fig. 74, it will be found, by placing the fore-finger of one hand on the projecting abdomen, and that of the other between the scapulæ, and exerting slight cross-pressure, that the patient will assume the erect posture shown in Fig. 79.

The same result can, in like manner, be obtained when there is a lateral tilt (*Vide* Fig. 78). To do this, place one finger under the right axilla, and another on the left hip, exert slight cross-pressure, and the patient will assume the improved position shown in Fig. 79. In both of the above

* Medical Society of the County of New York, Jan. 23, 1882.

[77]

instances the power, as will be seen, is applied external to the weakened muscles. As a permanent substitute for this finger-pressure, I have used, with entire satisfaction, the following appliance : In the center of a perforated strip of light tin about four or five inches long and three-eighths of an inch wide- (*Vide* Fig. 80), a short, thin piece of steel plate bearing a small ring, is soldered. During the application of the jacket these pieces are incorporated between the layers of bandage, the rings being allowed to project from the surface at the points indicated above for the application of finger-pressure. After sufficient time has been allowed for the " setting" of the gypsum, one end of a solid rubber cord, six or

Fig. 80.

seven millimeters in diameter, is passed up through the lower ring, and fastened to the upper. (*Vide* Fig. 78.) Fig. 81 illustrates the manner of securing the lower end of the elastic cord. A strip of non-elastic webbing, long enough to reach from the ankle to the knee, is sewed to the top of the shoe on the outer side. This is passed to the inside of the stocking, through a button-hole just above the upper margin of the shoe, as shown in the figure. The free end of this strip of webbing is provided with a ring. Through this ring the lower end of the elastic cord is passed and drawn sufficiently tense to overcome the distension, when it is fastened by a simple half-hitch, or knot. To diminish the projection of the belly, two elastic cords are necessary, one on each side. If, in addition, there is present a slight lateral tilt, it can be corrected by making greater tension with one or other of the elastic cords. This appliance can be used with any form of spinal support. In substituting the elastic cord for cross-finger-pressure, the power, as will be seen, is applied below the weakened muscles. I have had recourse to the same principle in correcting drooping of the head, due to weakness of the cervical muscles, occurring in the course of Pott's disease. The following case demonstrates the utility of this appliance. It will also serve to show how failures in treatment may result from an imperfect knowledge of mechanical therapeutics, or want of skill on the part of the surgeon in the application of agencies at his command.

The patient, a boy seven years old, having a good family history, first came under my observation Nov. 17, 1880. Until a year previous to that date he was healthy, active and unusually tall for one of his age. About this time he received a sufficiently severe injury to his spine to account for the sudden development of his trouble. Not long afterward angular curvature began to show itself, and a metallic spinal brace was applied. This was subsequently replaced at intervals with a plaster jacket, weighing, according to the father's statement, about four pounds. During the first application the patient was suspended mid-air by the head and axillæ with an improvised apparatus while an assistant, laying hold of his ankles, made forcible downward traction. With the patient suspended in this fashion, crying and struggling as best he could to free himself, the plaster bandage was applied. The duration of the operation, I am informed, was from half to three-quarters of an hour. Instead of a skin-fitting vest under the jacket, an ordinary loose undershirt, covered with a thin layer of cotton, was used. The shirt necessarily became wrinkled when the gypsum dressing was applied, and the cotton prevented the accurate adaptation of the bandage to the inequalities of the trunk. Hence any improvement in position which might have resulted from suspension properly practiced could not have been secured by a jacket so ill applied. In the application of subsequent jackets some improvements in apparatus were made, but the patient was always drawn up by his head and arm-pits so that he could not touch the floor with his toes, and he always cried. Each time, upon removal of the jacket, the skin over the projecting spine was found to be excoriated, and required treatment for several days before the application of another one. This treatment was continued nearly a year. The spinal deformity, as might have been supposed, rapidly increased.

Moving to New York, the father now consulted Dr. F. W. O'Brien, who upon examination of the case recommended that the patient be placed under my care, which was done on the above-named date. At that time the distortion had reached the unsightly degree represented in Figs. 76 and 77. Upon inquiring as to the patient's sufferings and clinical history, the father, naturally not a very acute observer, could give me almost no information. The patient, however, testified to no special pain or discomfort. He ate without complaint what he wished of food set before him, and no marked disturbance of sleep had been observed.

What a contrast are these statements to the picture actually presented to the immediate eyes of the observer (*Vide* Figs. 76 and 77), and how ineffectual are words to portray the details of that picture. Such a patient may complain very little, or not at all, and yet, an observing eye will not fail to detect the evidences of long continued suffering in the changed facial expression, the altered respiratory and cardiac movements, the conduct and condition of the extremities, and in fact, that of the whole body. A strong man may suppress outward manifestations of hunger when food is placed before him, but he can not prevent the watering of his mouth, or the secretion of gastric juice.

FIG. 81.

Fig. 76, although not showing all the details which could be wished, will serve to convey to the mind something of an idea of that pitiful expression of countenance which indicates hopelessness on the part of the little sufferer from ever escaping from the cause of pain. The forehead is transversely and perpendicularly furrowed and ridged, the corrugators and other facial muscles are tetanized, the nostrils dilated, the corners of the mouth drawn down, and there is a hopeless, semi-fixed gaze of the eyes. Upon being suspended my little patient expressed himself as feeling much better, though he had not previously complained. His respiration, which had been grunting in character, became more natural, and the expression of his face less anxious. On December 3d, 1881, I applied a plaster jacket over a skin-fitting knit shirt. During this operation two assistants were employed, one holding the shoulders and the other the hips of the patient in such a way as to correct as much as possible the lateral distortion and rotation of the spine. By this means and the suspension of the patient, the deformity was at once very considerably ameliorated, and an accurately fitting jacket being applied, approximately secured the patient in the improved position. Three months later, upon removal, this jacket was found to weigh twenty ounces.

Within the following ten months I saw the patient three times, and applied as many new jackets after the plan above described. At the end of this time, during suspension, no rotation or lateral deviation of the spine was observable. However, as I have not infrequently noticed in other cases, so in this, notwithstanding there was no deviation of the spine during suspension, and no deficiency of skill in the application of the jacket, a lateral tilt (*Vide* Fig. 78), showed itself when the patient was allowed to stand upon his feet, owing to the greater weakness of the lumbar muscles upon one side of the vertebræ. To overcome this I made use of a solid elastic cord applied in the manner I have already

[79]

ILLUSTRATED MEDICINE AND SURGERY.

described. Fig. 78 shows to what extent the patient leaned to one side after a careful application of the jacket. Fig. 79, from a photograph taken ten minutes afterward and with the elastic cord tightened, shows how effectually it corrected the lateral distortion due to muscular weakness. It will be seen that the elastic cord does not restrict the movements of the patient. After stooping, or bending to one side, the patient, instead of resuming the faulty position shown in Fig. 78, remains as in Fig. 79. This simple mechanical adjuvant to the plaster jacket proved a complete success, not only in its immediate but its permanent effects. Some two months afterward the patient returned to my office, and had recovered so much strength in the weakened muscles, by their daily exercise, that he was able, without the aid of the jacket, or other artificial support, to stand quite as erect, as in Fig. 79.

[80]

TWO LARGE TUMORS OF THE FEMALE BREAST.

BY WILLIAM T. BULL, M.D.,

Surgeon to St. Luke's and Chambers Street Hospitals, New York.

I. Spindle-celled sarcoma of the female breast. Removal followed by recurrence and death from exhaustion. Secondary deposits in the brain, lungs and retro-peritoneal glands.

M. H., a widow, æt. 42, a native of Ireland, has always been in good health till her present trouble, and has never suffered any injury. No children. Four years ago a hard nodule appeared just above the nipple of the left breast. It was the size of a marble, and painless. The menses ceased about this time. The lump increased in size slowly, and up to nine months ago it had reached half its present dimensions without giving rise to annoyance other than from its weight. Since then, however, it has grown rapidly, and been very painful at times, and she has lost appetite, flesh and strength. Sloughs appeared on the surface of the tumor after applying poultices, and a highly offensive, thin discharge issued from beneath them. To obtain subsistence she begged in the streets, exhibiting the growth to passers-by. Finally, unable to obtain food, she applied to the Chambers Street Hospital, in a condition of extreme exhaustion, and after two days' care was transferred to the New York Hospital. On admission she is extremely emaciated and weak; features pinched, eyes sunken, stomach irritable, and rejecting all but small quantities of fluid diet. P. 104, weak; R. 32; T. 101.6. The tumor occupies the left mammary region, and is half as large again as the woman's head. It is smallest at its base, and freely movable on the chest-walls. The surface is smooth, but uneven, hard in some parts, soft in others, and the skin covering it movable, thinned, traversed by large veins, and presents three superficial sloughs, each two inches in diameter, which discharge a thin, stinking pus. No axillary glands to be felt. Heart and lungs normal.

The following day she was somewhat stronger, but the evening temperature was 103°; P. 106. Urine, sp. gr. 1020, no albumen. No evidence of internal deposits. Stimulants, quinine and fluid nourishment. Moderate pain, occasional vomiting.

On the third day the sloughing of the surface had extended, the stench was overpowering, and the discharge more profuse. The ulcers were disinfected with Labarraque's solution, and the atmosphere of the room saturated with spray of carbolic acid. General condition a trifle better. After consultation with Dr. T. M. Markoe it was decided to operate.

The fourth day after admission the tumor was removed under ether. An assistant held it up vertically, and two curvilinear incisions were made on each side of its base through healthy but thinned integument. The dissection was done rapidly, and with almost no loss of blood, an assistant grasping each bleeding point with forcipressure forceps. A small portion of the sheath of the pectoralis major muscle being opaque, was removed with scissors. Eighteen carbolized catgut ligatures were applied, wound united with silk sutures, two rubber drains being left at each extremity. All antiseptic precautions were adopted save the spray, and a Lister dressing used. The woman was but slightly depressed by the operation.

There was primary union at a few points only, and the wound, remaining aseptic, healed by granulation in four weeks. But the general condition was very bad. The fever was very moderate, but there was great depression, with irritable stomach and an obstinate diarrhœa. This gradually stopped, and at the ninth week after admission she was discharged in greatly improved condition, but still anæmic and thin. The cicatrix appeared healthy.

The tumor weighed eight pounds. Fully one-third of its mass was softened and necrotic. Sections from different parts presented the same appearance under the microscope, that of closely-aggregated spindle-cells, the intercellular substance being homogeneous in character, with numerous large blood-vessels filled with blood. After leaving hospital the woman improved in health and was able to work.

Two months later a nodule appeared in the cicatrix, which in three months more grew to the size of a child's head, with constant pain, and ulcerated in several places. It was movable, however, on the walls of the chest, and the axilla was free from enlarged glands. The woman's general condition was as bad as before the first operation. The growth was removed by Dr. Wise, leaving an open wound. Death ensued four days later from exhaustion.

Autopsy (by Dr. George L. Peabody, Pathologist to the New York Hospital). Body emaciated. Rigor mortis well marked. Œdema of both lower extremities. A large wound, seven inches in longest diameter, over site of left breast. Peritoneum: no fluid in cavity, no signs of peritonitis. Diaphragm: right side at fourth rib, left side fourth intercostal space, old adhesions at left apex. Lungs: two small nodules (1¼ and ½ inches in diameter) between apex and lower border of upper lobe of left lung. Right lung presents several similar nodules. They vary in size from that of a pea to that of a horse-chestnut, are white, and of soft consistence. Heart: pericardial thickening anteriorly and posteriorly, firm clots in all the cavities. Aorta and mitral valve atheromatous. All valves sufficient. Muscular tissue brownish and flabby. The spleen rests on a hard, firm retro-peritoneal tumor, which occupies the left hypochondrium, and is overlain by the stomach and pancreas. The mass is hard, friable, and globular, and weighs 800 grammes. Four inches below this is another tumor, of 300 grammes weight. The left kidney contains several whitish deposits in one pyramid, surrounded by zones of congestion, anæmic and normal. Right kidney likewise normal. Stomach small and anæmic —its mucous membrane normal. Duodenum, intestines and liver normal. Brain is anæmic. In the extreme end of left occipital lobe a cavity just within the limit of gray-matter, containing a soft mass as large as a horse-chestnut. It is whitish, and shells out of the cavity easily.

Microscopic examination. The growths in the lung consist of cheesy masses, in whose borders are masses of round and spindle-cells, lying in close contact with one another in places, and again with slight amount of intercellular connecting tissue. Tumors behind the peritoneum are of similar structure. The tumor of the brain is a round and spindle-celled sarcoma, as is also the tumor removed from the chest wall.

II. Fibroma of the female breast (Fig. 83). Removal. Recovery.

B. L., æt. 40, a strong, healthy Irish woman, was admitted to the New York Hospital July 30, 1879. Married, but no children. Six years ago was kicked in the right breast, and had an inflammatory swelling there, which subsided without suppuration in a few weeks. A year later the breast began to enlarge, and has continued to do so till it is now as large as an adult head. She has had no pain in it, and has been inconvenienced only by its increased weight and bulk.

On admission the right breast is as large as her head, the skin over it normal but thinned, and traversed by large veins. Nipple likewise normal. The surface of the tumor is irregular, but smooth and rounded, its outlines sharply defined, its consistence firm. It is not tender, and it is freely movable on the subjacent structures. No glands to be felt in the axilla.

The tumor was removed August 2, 1879, antiseptically, but without spray. Two curved incisions including an elliptical portion of integument, were made down to its capsule, and the mass readily enucleated. The sheath of the pectoral muscle was healthy. No large blood-vessels.

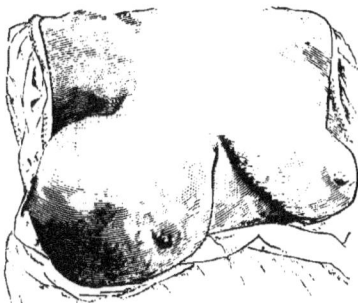

Fig. 83.

Catgut ligatures and silk sutures, two drains and a Lister dressing. Two dressings were applied in the following ten days, when the wound was completely closed except at the points of exit of the drains. No fever. Bals. Peru. Completely healed on the sixteenth day, and the woman discharged.

The weight of the tumor was 3¼ pounds. Its cut surface was firm and elastic, whitish in color, and yielded no juice on squeezing or scraping. Sections from various parts showed under the microscope a few spindle-cells thinly scattered in a stroma of well-defined fibrous tissue. Some dilated milk ducts were to be seen, and a few shrunken acini ; very few blood-vessels. The great mass of each section is made up of fibrous tissue.

The woman promised faithfully to report in case of recurrence. All efforts to find her now (three years having elapsed) are fruitless. But from the size of the growth, it is possible that a local recurrence has taken place. Gross states this to have been the case in " rather more than once in every sixteen cases." The size of the tumor is its remarkable feature, since Billroth mentions one as large as a goose-egg as being of exceptional proportions. But Gross quotes one case in which the tumor weighed upward of twelve pounds. The rate of increase was unusually rapid in this case, too, since it presented no sign of cystic degeneration, which has been observed in the more rapidly growing tumors.

Case I. illustrates the malignant character of the spindle-celled form of sarcoma—a feature which has been denied by some authors, but which has been definitely established by the researches of Prof. S. W. Gross. In eleven out of sixteen cases "there was local or general occurrence, which appeared, on an average, in thirteen months." In this instance the duration of life may be roughly stated as four and a half years. The time at which the internal deposits occurred is uncertain, but, from the extreme prostration, it is probable that they existed before the first operation was performed. The freedom from disease of the axillary glands is a well-known feature of this sort of neoplasm.

SARCOMA OF THE PHARYNX AND SECONDARILY OF THE NECK.

BY L. A. STIMSON, M.D.,

Professor of Surgical Pathology, University Medical College, New York; Surgeon to the Presbyterian and Bellevue Hospitals, New York.

Otto B., ten years old, was admitted to Bellevue Hospital in the Fall of 1879, to be treated for a tumor of the pharynx and neck, which had made its appearance the previous April. The interference of the growth with respiration made tracheotomy necessary in November. He came under my care January 1, 1880, and I was urged by the boy's parents to attempt the removal of the tumor. The following notes were taken at the time: "The boy is pallid but fat; the cavity of the mouth occupied by a tumor of irregular surface, at first congested in appearance, but now paler, which is attached upon the left side in the tonsillar region, the attachment extending down the inner side of the lower jaw, as far forward as the second molar tooth. The finger cannot be passed beyond it to determine the posterior limits of the attachment. The left side of the neck is swollen by a mass which extends from below the ear obliquely forward and downward past the angle of the jaw. This mass is composed of at least two distinct parts, the upper and larger one giving a sensation of fluctuation, the lower one being harder, and feeling like an enlarged gland; the surface of the larger one is smooth and uniform, the skin normal and non-adherent. Patient bleeds a little from the mouth every night; breath fetid by sloughing of the buccal tumor on its under side. The tumors in the neck are apparently not continuous with the one in the mouth.

"Feb. 5.—The mouth was laid open by an incision carried from its angle to the angle of the jaw on the left side; I could then pass the finger around the tumor, and found it deeply ulcerated in its under surface, and attached to the posterior pillar of the fauces, as far out as the uvula; apparently it did not extend back of the posterior pillar. I passed a wire around it and cut off as much as possible with the écraseur, and then removed more by tearing and cutting with scissors. I was prepared to divide the jaw behind the teeth, but refrained because of the apparent impossibility of making a complete removal of the growth, which proved to extend down the side of the pharynx, behind the tongue. The portion that was left was crushed between the fingers, in the hope that it would slough out. The part removed was pale, yellowish white, very soft and juicy.

"The incision through the skin was then extended towards the ear, and both tumors removed by careful dissection. The larger was 3 inches long, 1¼ inches in diameter, and lay upon the carotid, and in front of, and partly under the sterno-cleido mastoid.

"Microscopic examination shows the mass to be composed of small, round, nucleated cells, having no alveolar arrangement, and with, in some places, an abundant intercellular substance. No sign of mucous tissue. Prognosis bad. Feb. 6.—Pt. is comfortable. Temp. 99°.

[85]

"Feb. 8.—Pt. is sitting up. Breathes freely through his mouth.

"Feb. 21.— The incision has healed within the mouth, and, on the outside, as far back as to the angle of the jaw; behind that is a cavity, which is granulating. The tracheotomy tube was removed a few days ago, and the opening has closed.

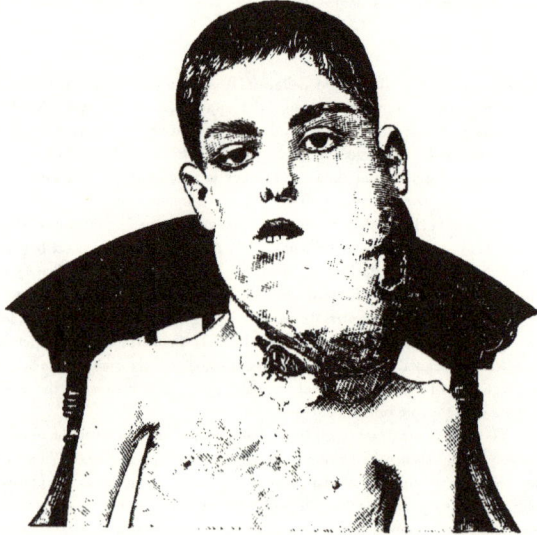

FIG. 81.

"Mar. 3. —The tumor is growing rapidly in the soft palate, and a gland above the left cheek has become as large as a horse-chestnut.

"Apr. 13.— Replaced the tracheotomy tube; the trachea was displaced to the right about an inch. The growth has extended down the pharynx beyond the reach of the finger, and involves the margin of the glottis.

"May 19. —The tumor in the neck is enormous, half as large as the boy's head. The glands on the opposite side are involved.

"July 12.— Otto died to-day of exhaustion. The tumor had grown to an enormous size, and a week before his death ulcerated on the cheek, the ulcer extending through into the mouth. The trachea and the tube were pushed far over to the right. The veins on the front of the chest were much enlarged. Towards the last he had some difficulty in swallowing even liquid food."

No autopsy allowed. The photograph was taken two or three days before his death.

NO. DOUBLE EQUINOXARUS.
(Case of Prof. Louis A. Sayre.)

DOUBLE EQUINO-VARUS.

RESULT OF INFLAMMATION; SUBCUTANEOUS TENOTOMY OF TENDO ACHILLIS
AND PLANTAR FASCIA OF EACH FOOT; IMMEDIATE REPLACEMENT AND
RETENTION IN NORMAL POSITION. RESULT—NEARLY PERFECT RESTORATION
OF FORM AND MOTION IN SIX WEEKS

BY LEWIS A. SAYRE, M.D.,

Professor of Orthopædic Surgery Bellevue Hospital Medical College; Consulting Surgeon Bellevue Hospital, &c. &c.

Miss M. de O———, aged 14½, of Colombia, South America, was brought to me by Dr. Forero, of that city, on the 20th of May, 1882, suffering from double club-foot—Equino-Varus in nearly its most exaggerated form.

Her parents were perfectly healthy, and she had always been healthy until 10 years of age, when, after exposure to cold while in a state of perspiration, she was seized with an attack of acute rheumatic fever, involving the knees, ankles, wrists and elbows of both sides, the orbicular joints not being attacked.

She was confined to her bed for several months. As she recovered from the acute pain, the heels began to draw up, and the feet became inverted, as seen in the accompanying Plate XIV. She was unable to stand without support, and had been unable to walk without crutches for the past four years. Her body was unusually large, but the lower extremities were badly developed, and from the knees down the limbs were atrophied, very cold, and quite purple.

The feet could not be extended, nor the heels brought down by the strongest manipulation, and when pressure was made on the tendo-achillis or plantar fascia, while thus stretched, it was followed instantly by a severe reflex spasm, showing that these tissues had become structurally shortened, and therefore required section or division before the feet could be restored to their proper position.

The following is a very important rule in practice, and I believe I was the first to establish it. When any contracted tissue, whether muscle, tendon or fascia, is stretched to its utmost limit, and pressure made by the tip of the finger or thumb on this tissue, produces a reflex spasm, it must be divided before any further elongation can be obtained. On the contrary, if, when the contracted parts are thus stretched, this kind of pressure does not produce a reflex spasm, those tissues can be

[87]

elongated to their normal length by the constant use of an elastic tractile force, and therefore do not need division.

I published this law many years since in the second edition of my Manual on Club-foot, and a very extensive practical application of the principle during the past ten years has proved to me its value.

In this case the law was clearly exemplified, for she had applied every device invented for the relief of club-foot for three years without the slightest improvement in the direction of her feet; indeed, an increase of the deformity by the callosities produced by pressure over the bony prominences had taken place. These finally became so painful that all treatment had been abandoned for the past year, and she depended altogether upon her crutches for locomotion. The callosities had all subsided, and her feet were therefore in good condition for operation.

I operated on the 23d of May, 1882, assisted by Dr. Forero, of Colombia, S. A., my son, Dr. Lewis Hall Sayre and Dr. Robert Taylor. After the patient was fully under the influence of chloroform, I divided subcutaneously the tendo Achillis and plantar fascia of each foot with the loss of only a drop or two of blood, closed the wounds with adhesive plaster, covered the foot and ankle with a thick layer of cotton wool, which was secured by a roller bandage, and then, by the application of some considerable manual force I brought the feet *immediately* into their normal position, and retained them there by the application of my foot-board, adhesive plaster and a roller-bandage.*

The entire dressing was perfectly completed while the patient was still under the influence of the anæsthetic. No constitutional trouble followed, and at the end of eleven days the dressings were removed for the first time, all the wounds were perfectly united, without the formation of a drop of pus.

The new tissue between the severed ends of the Achilles tendons (more than an inch in length) had already become so firmly organized that very slight movements of the heels were quite perceptible when the patient made voluntary contractions of her gastrocnemii muscles, showing that adhesion had already taken place.

The feet were again dressed, as after the operation, for another week, when all dressings were removed, and the feet and limbs treated every day by massage and passive movements for half an hour, and the application of the Farradaic current for five minutes.

Four weeks from the date of the operation she began to walk with the aid of a "Hudson's Shoe,"† see Plate XIV.

Daily manipulations and the application of electricity were continued until July 5th, when she could walk without any support, and the muscles of the leg had become quite prominently developed, see Plate XIV.

This young lady called on me August 14th, eleven weeks after the operation—had walked over two miles that morning, and the increase in the size of the muscles of the legs was astonishing. She still continued the massage and electricity.

* See description in author's Manual of Club-foot, page 39.
† For descriptions see author's Manual Club-foot, 4th ed.

EXTRANEOUS MICROSCOPIC MATTER IN ANIMAL FLUIDS.

BY E. P. BREWER, M.D., PH.D.,

Of Norwich, Conn., Late Surgeon to Hartford Hospital, Conn.

In the microscopy of solids little difficulty is experienced from the admixture of foreign matter, but in the examination of fluids where its features are made up of a large number of suspended elements, and, further, is exposed to contamination by deposits from the air, and in the transfer from vessel to vessel, we frequently meet with diverse and perplexing substances, calling forth considerable labor for their differentiation from the correct elements of our specimen. When encompassed by these difficulties our enthusiasm, I fear, sometimes falters, and we regard microscopy, in ease of acquisition at least, as Huxley did Bathybias, failing to fulfill the promise of its youth; however, with perseverance the shadows fall behind, and are forgotten in the advance of the bright light of victory.

The drawings of Fig. 88 represent the most common of extraneous matter met with in fluids. Some of them, as cotton (*b*) and woollen fibre, (*c*) are almost constantly present, and thus become familiar to the microscopist in the early part of his career. Their characters are so decided that they seldom give the least difficulty in identity. Cotton fibres (*b*) are generally long, much twisted, and marked with lines confusingly interlacing, so that no definite repetition of structure is apparent; the diameter is irregular, both in the same and different specimens, though generally large; perhaps, in their extremes, from $\frac{1}{x}$ to $\frac{1}{xx}$: in the ends are frayed out and the fibres discordantly arranged. Woollen fibre (*c*) presents a striking contrast with cotton fibre. Its dull, pearly lustre, distinct, scale-like

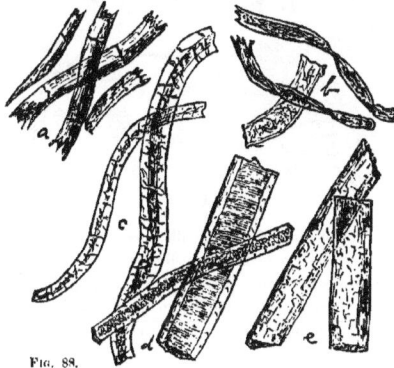

FIG. 88.

markings, long, graceful curves, large diameter, and nearly square broken ends, separate it widely from all other extraneous matter. Flax fibre (*a*) will seldom give trouble from its resemblance to intrinsic products. The fibres are straight, or in short curves, and transversely marked at regular intervals. The ends are fibrillated and the fibrillæ of nearly an equal length. Although slightly resembling a tube-cast, the large diameter (not varying much from $\frac{1}{50}$ of an inch) will alone sufficiently serve to distinguish them. Like cotton or woollen fibre, the color will vary with its source. Not unfrequently cat-hairs (*d*) beget annoyance, especially when of *small* diameter and after long maceration. As depicted in its normal condition, we note a granular centre enclosed in transparent walls. On prolonged soaking in water the granular centre becomes less defined, and if the hair

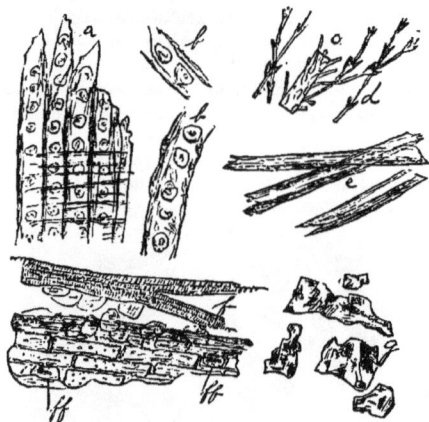

FIG. 89.

be, as it may, the diameter of a tube-cast, and the length also, in conformity, the appearance honestly apes a granular tube-cast, and, unless we observe with some care, our conclusion will be erroneous. On careful scrutiny, the ends of the suspected cast will be found distinct and irregular — commonly crenate — with a cup-shaped depression in the end, probably formed by the separation of the granular matter, and leaving the wall projecting. Adding acetic acid (glacial) the object will become homogeneous under an objective of $\frac{1}{4}$ inch, but will again betray the original characteristic structure on the addition of the carmine solution; however, as a rule, the length of the specimen, and the progressive tapering, preclude a possibility of mistake. Cat-hairs of a large diameter (*d*) markedly differ from the small hair. In the place of a granular centre there are distinct, closely placed transverse striæ, limited on either side by a homogeneous wall. In all specimens the ends are non-fibrous. Human hair (*e*) is exceedingly easy of identity. Its visibility to the unaided eye, length, distorted ends with the short fibrillæ, link as adequate evidence for the recognition of the object and to exclude the possibility of its being a cast.

Fibres of silk, Fig. 89 (*c*), particularly if uncolored, commonly approach more nearly the tube-cast, *in size*, than any of the preceding objects, and indeed are closely allied in appearance to some specimens of hyaline casts. The diagnosis must mainly rest upon their smooth, clean cut, and glistening appearance, fibrous extremities, the regularity in the relative size of the specimens (if more than one be present), the small diameters, and the rapidity of the staining with carmine. Portions of feathers (*e, d*) occasionally find access to the sputa and urine, but as they bear no apt resemblance to normal or pathological ingredients, they require no special treatment. The straight, barbed fila are generally separated from their shaft and are numerous. Although they have been confounded with nerve tubes, no rational excuse will exculpate the microscopist for the error. The regular barbs on the secondary shaft and the fine filament emanating from the terminal barb amply prove the specimen to be feather. Mineral matter (*g*) is also frequently present. Their presence largely depends upon the deposit of dust. In size and contour they are extremely variable, some particles are large, others small, some rough and ragged, others smooth and polished. I have sometimes thought it possible to learn of the locality of the patients' dwelling simply by means of the lineaments of these mineral particles. When rough, angular, and slightly translucent, I intuitively associate with them a dry, sandy locality; when more regular, and with smooth, polished corners, the action of attrition in water is probable; opaque particles of conglomerate character indicate damp, low land, while rough opaque granules may fairly be recognized as an attribute of ordinary dry soil. Translucent particles of silica are occasionally confounded with epithelial debris, but may be accurately distinguished by their imperviousness to coloring matter.

The fibres of coniferous wood (*a*, *b*) are the best calculated to deceive of any extraneous matter ever present in the urine. By manipulation with needles fixed in a wooden handle, the fibres become separated and disseminated in the fluids under inspection. In water they become soft and swollen, and may appear like tube-casts. When the outline and size are in common with tube-casts, the pores (well shown in *b*) decidedly resemble the epithelial cells in casts. When large masses are present (*b*) the peculiar conformation of coniferous wood is competent for distinction, but in isolated fibres containing only two or three pores, and these more or less distorted by the swelling of the fibre, identity is difficult, and the confounding of the latter with a tube-cast not infrequent. If we view the fibre—the suspected cast—with care it will be noticed that the fibre and epithelium-like cells are more distinct than in a true cast of its type, the supposed nucleus in the centre of the cell is unduly large and without a nucleolus, while carmine rapidly stains the whole substance except the nucleus. Acetic acid notes no specific actions. If the definite of the object-glass be good, the suspected nucleus may be resolved into a pit in the cell. Occasionally the ends will have a jagged outline, which will assist in diagnosis. Dr. Richardson (Med. Microscopy, 1871) suggests a rule which, in some doubtful cases, may be of assistance. He advised that the thin cover glass be firmly pressed with a mounted needle over the suspected body, as seen with a half-inch objective, when, if a tube-cast, it will be crushed beneath the force applied, whilst the cover glass would probably break before a hair, or a body like wood fibre would be more than slightly flattened. The writer remembers a ludicrous error, created by one of these fibres, which has since served to emphasize the dangers of mistaken identity of these objects, and also to fix in mind the specific characters of the deal fibre. A microscopist of considerable experience, while examining some vomit to determine if it contained evidences of malignant disease, encountered, as the only evidence of disease, a coniferous fibre, which he construed to resemble an exfoliated fragment of the gastric mucous membrane. The structure of the specimen was fully appreciated, but misinterpreted. The pores were thought to represent the openings of the gastric follicles, while the peculiar cellular structure was believed to be the result of long maceration and pathological variation. The specimen was exhibited to a microscopical society of one of our large cities, was duly examined and discussed, yet its real nature was not disclosed, and, in fact, not known by its discoverer until some months later.

Some months ago, in examining sputa from a case of laryngeal phthisis, in which the ulceration apparently descended into the muscular tissue of the larynx, I found an object which is well depicted in the upper segment of Fig. *f.* The close resemblance that it bore to muscular tissue, led me to carefully examine it. The carmine solution was quickly absorbed, and rendered the structure quite distinct. The general appearance was not unlike striated muscular fibre, but from the ends fine, curled, thread-like projections were seen, which were recognized to be the spiral vessels of a vegetable growth. This at once dissipated all doubt of its origin, although its source remained a mystery. Other specimens reflected a structure as shown in the lower part of Fig. *f.* The cells between the upper and lower segment are the cells of vegetable parenchyma; the oblong cells below, epidermic tissue; and the cells to the right and left marked double f (*ff*), stomata, or breathing pores, in the epidermis. The whole proved to be finely comminuted straw from the patient's bed.

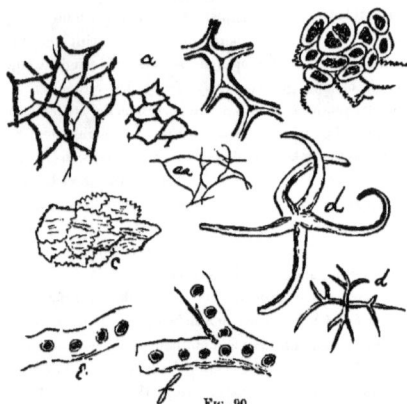

Fig. 90.

Fig. 90.—Various vegetable products enter into the composition of animal fluids in a partially disintegrated condition. This is most frequent in saliva and sputa, having origin in the ingesta. A few of the forms of partially disintegrated vegetable tissue (*a*) found chiefly in saliva, resemble pavement epithelium, but may be differentiated by the fibro-cellular bands, the absence of alveolar tissue, and the non-action of liquor potassæ on the tissue. This, and also leptothrix buccalis (*a, a*) are distinguished from pulmonary elastic tissue by the variance of their diameters, and also by the condition of the ends of the fibres. In (*a*) the diameter is greater and the ends *frayed out;* in (*a, a*) the fibre is finer and its ends pointed. Pulmonary tissue, it will be remembered, is possessed of square extremities. Leaves of tea (*b*) are found in saliva, and not rarely in the urine, with a mission of deception which a hysterical patient prides herself is "such an easy matter." The cells are known by the large mass of dark granular matter which they contain, and the association with them of the fibro-vascular filaments. Isolated epidermis (*c*) from plants is known from epithelial cells by the absence of a distinct nucleus, their angular outline and irregular margins. Coloring matter is rapidly diffused through the tissue without developing new traits of structure. In the seasons of foliage, many forms of vegetable hairs separate from plants and float into our fluids under examination. The types are varying, but from the buoyancy of the stellate (*d*) type they become the most common form met with. Although they are not to be confounded with any other ingredient found in animal fluids, they oftentimes puzzle one on their initial introduction. Nitric acid increases their transparency, but elicits no distinct structure. Fresh water algæ exceptionally engender difficulty by their assimilation to tube-casts. I have met two varieties. The first was a hyaline alga which would be well presented if the cells were obliterated from *c*. It had a diameter of $\frac{1}{15}$th of an inch, and responded affirmatively to the pressure test ; however, on careful focusing, I discerned an empty cell, the specimen evidently representing an alga with the endochrome expunged. The second alga is shown in *c* and *f*. The specimen *e* was first noticed, and its semblance to a tube-cast is sharply drawn ; here, also, the pressure test gave affirmative results. It will be noticed that the cells are granular, while the other portions of the specimen are hyaline ; this state excited suspicion and led to further search, which resulted in the finding of *f*. In that the cells were green and the structure branching, which, of course, destroyed the possibility of its being a cast, and simultaneously identified it as Homospora mutabilis, one of the Palmellaceæ.

I afterward discovered that I introduced these products into my specimen on my dipping-rod, which was cleansed in water containing them.

Fig. 91. — Oil (*a*) is known by its high refractory power, its globular shape and single, narrow, dark border. This border-line distinguishes it from air-bubbles, in which three well defined rings are seen,—a light interior and exterior ring, with a dark intermediate ring. In urine, oil may be diagnosticated from chyluria by the large size of the globules and from milk (*g*), which is sometimes added by dishonest or hysterical people by the absence of colostrum corpuscles (*g, g*). Muscular tissue (*e*) is greatly

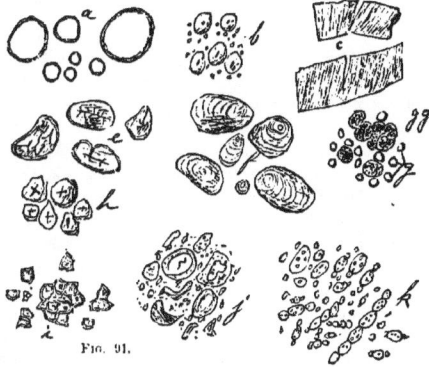

FIG. 91.

changed by soaking in fluids ; nevertheless, it sufficiently maintains its original character for recognition. In the application of mustard poultices for counter irritation, oftentimes a portion will escape and dry, and then be scattered in the urine or sputa as dust (*b*). This occurred in a case under my observation. The general aspect of mustard is allied to arrowroot ; the starch granules are about the same size and appearance ; again, potato starch (*f*), bean starch (*c*), maize (*h*), rice (*i*) and baked wheaten bread (*j*) may find access to the saliva and urine. In the last object (*j*) the starch granules are broken, and changed almost beyond identity by baking. All varieties of starch are quickly identified by the iodine test. A drop of the tincture may be applied to the stage and submitted to a gentle heat, and if starch be present the granules will become purple or blue. The addition of yeast to the urine is betrayed by yeast cells (*k*) in different stages of multiplication.

With the practitioner of general medicine, little or no diagnostic or prognostic value is imposed upon the microscope beyond the recognition of the characters of urinary tube-casts ; hence it is that he should be conversant with the contaminating substances standing in relation with or resembling these casts. We have cited sufficiently in detail, I believe, the guiding features of these substances to lend a helping influence, and will now append a few general laws which will materially assist in reaching the truth when we meet with suspicions and ill-defined specimens.

The distrust which exists among some physicians of the reliability of the microscope, even in the prognosis of Bright's disease, is I fear, directly due to their own hasty conduct in basing an opinion upon the character of one, two, or four casts. This is certainly the most fertile source of error in microscopy, for I have known diagnoses of Bright's disease (chronic) to rest solely upon the presence of three or four granular tube-casts and general debility. And in those cases of this class that I have followed, no ill effect has ever developed, although the casts were present from time to time. If we have urine from a case of indubitable disease of the kidney and are determined to classify its type and stage, we never base our opinion on a few casts, two, three, or four of a particular kind, but should examine a large number, and direct our attention and opinions to the general character of the deposit. Thus, in acute Bright's disease, a few casts will contain oil, but this will not lead us into error, for experience has taught us of their frequent presence, followed by total, rapid, and complete recovery. Why, then, should we desert our guiding star, and diagnosticate disease from a single or a doubtful cast, and upon evidence almost unsupported by general symptoms, when we refuse to diagnosticate the stage of a typical case of the same disease, on ten times that number of casts. Therefore, let it be a law *never to base an opinion upon a few casts.* With this caution in mind, the presence of a few contaminating, imitative types will never confuse or lead us into error.

Tube-casts vary much in size, according to their origin and manner of separation. From tubes of equal diameters the casts may vary in size; if the epithelial layer be of ordinary thickness, a cast of medium size will form; if it be swollen, the cast is small, or if the epithelium be stripped off, the cast will be large; consequently in the same specimen of urine the extremes in size may be seen varying from $\frac{1}{250}$ to $\frac{1}{1000}$th of an inch in diameter. This irregularity is the most constant feature in casts, and should have due consideration in reaching our conclusions. With few exceptions, extraneous specimens will be of the same diameter or nearly so, and never, to my knowledge, show the trasitional stages between the extreme diameter of $\frac{1}{250}$ and $\frac{1}{1000}$th of an inch, as noticed in casts. We, therefore, declare that *irregularity in the relative diameters of casts is of value in their diagnosis.*

The intensity of illumination may also aid in complex cases. With the student, the detection of hyaline, or finely granular casts is a difficult task, for with bright illumination he seeks the subtle object, and unless some friendly, coarsely granular cast falls in the range of vision, his search will be fruitless; later in his career he learns to seek with dull illumination, and always to good advantage. Comparing the requisite intensity of illumination for the clear discrimination of objects, a natural classification is made particularly of true and false casts. *True casts may be best seen with dull illumination,* especially the hyaline types, yet, it remains true through the whole series; *false, imitative casts are best seen with full illumination;* most assuredly, they may be seen with dull illumination, but not so clearly; on the contrary, the true cast is often invisible with bright illumination.

SYPHILITIC ULCERATION IN THE UPPER AIR PASSAGES.

BY CARL SEILER, M.D.,

Lecturer on Diseases of the Throat at the University of Pennsylvania, &c., &c.

Annie A., æt. 18, waitress, a healthy-looking girl, complained that her speech was almost unintelligible, her voice having a very nasal sound and of the character noticed in cases of cleft palate. She stated that when she was six years of age she had had sore throat for several months. Since then she had enjoyed good health, but her speech had gradually become worse.

On examination I found the following condition of the palate, pharynx and larynx: Both anterior and posterior pillars on the right side had disappeared, as well as the tonsil, while the margin of the velum about one-half inch from the base of the uvula was tightly adherent to the wall of the pharynx at the median line. From the point of adherence a fold of mucous membrane ran down toward the base of the tongue, and was evidently the remains of the posterior pillar. The uvula was drawn over toward the left side and hung in a slanting position. The tonsil and both pillars on the left side were normal. A short distance above the root of the uvula was a stellate cicatrix. The velum and uvula were but slightly movable, and did not close the naso-pharyngeal opening during phonation. This explained the character of the voice. A rhinoscopic examination showed the right side of the vault of the pharynx to be closed by cicatricial tissue, while the left side and the posterior nares appeared normal.

Although vocalization was normal, and the patient complained of no symptoms pointing to disease of the larynx, I made a laryngoscopic examination and found the following lesions:

The epiglottis had been destroyed by ulceration down to the level of the glosso epiglottic sulcus on the left side, and projected but little above that level on the right. Its margin was ragged and of a glistening white appearance showing the line of mucous membrane on the other side sharply defined. The left ary-epiglottic ligament with the cartilages of Wrisberg and of Santorini had also been destroyed, and in their place was a band of cicatricial tissue making a deep excavation in the outline of the laryngeal image on the left side. The left arytenoid cartilage was represented by a slight elevation under the mucous membrane, out of which projected a small spicule, evidently the remains of the apex of the arytenoid cartilage. The ventricular bands as well as the vocal cords were normal and their motion during phonation perfect.

[95]

An operation with a view to loosen the velum from the pharynx was deemed impracticable. There is no doubt in my mind that these cicatrices were the result of syphilitic ulcerations, the disease having in all probability been inherited.

The points of interest in this case are the absence of all laryngeal symptoms, and particularly the fact that deglutition was normal although the projecting portion of the epiglottis was absent, and that phonation irrespective of articulation was good in spite of the loss of the greater portion of the left arytenoid cartilage.

Case II.—John J., æt. 23, machinist, consulted me in regard to a severe cold in the head which had lasted several weeks. I recognized in him a patient who had been under treatment at the University Throat Dispensary three years before for secondary manifestations in the larynx. He told me that after his discharge from the dispensary he had continued his medicine (iodide of potassium) for several weeks only.

Upon examination of the anterior nares I found a perforation of the cartilaginous septum by ulceration, which latter extended along the floor of the right nostril and invaded the lower turbinated bone. The left nostril was free from ulceration, but obstructed by a large anterior hypertrophy of the tissue covering the lower turbinated bone. Inspection of the mouth showed a small ulcer on the palate situated in the median line and about three-quarters of an inch from the teeth. On the pharynx a gummatous swelling occupying nearly the whole of the right side was seen, and the rhinoscopic mirror showed it to extend high up into the vault of the pharynx. Examination with a probe showed that a communication existed between the ulcer on the floor of the nose and that on the hard palate, and demonstrated the existence of necrosed bone.

The patient was placed under specific treatment, and the ulcerations in the nose, after having been cleansed from their secretions, were thoroughly cauterized with the galvano-cautery, after which daily applications of a solution of acid nitrate of mercury were made to their surfaces. In a few weeks the ulcerations began to heal. A sequestrum in the palate became detached and was removed through the nose, leaving a small oval perforation in the palate. About that time the gummatous swelling in the pharynx commenced to ulcerate in its entire extent, in spite of the vigorous systemic treatment, and this ulcer was also cauterized with the galvano-cautery and the acid nitrate of mercury. Several weeks elapsed, however, before it showed any signs of healing, and then was very slow in disappearing, so that the treatment extended over several months. Throughout the duration of the pharyngeal ulceration the patient experienced great difficulty in deglutition.

At the present time, five months after the healing of the pharyngeal ulcer, no cicatricial contraction has caused deformity, and the scar is barely visible. The right nostril also appears normal with the exception of the perforation.

It is my experience that ulcerating gummata and deep syphilitic ulcers of the mucous membrane are always due to the breaking down of gummata, and if treated locally, the destruction of tissue can be confined to the swelling itself. It is the extension of the ulcerative process beyond the limits of the gumma which gives rise to destruction of tissue and cicatricial contraction.

This is illustrated by my first case, in which the ulcers were not treated locally, and consequently a large amount of tissue was destroyed.

LARYNGO-TRACHEAL DIPHTHERIA IN AN ADULT.
(Case of Dr. W. S. Cheesman.)

LARYNGO-TRACHEAL DIPHTHERIA IN AN ADULT.

BY WM. S. CHEESMAN, M. D.,

Late House Physician, Bellevue Hospital, New York.

On the 1st of March, 1882, I was called to see a gentleman aged thirty-two, of great muscular and constitutional vigor. He had been exposed in a storm some few days before, and was suffering from what he supposed to be a severe cold. There was much soreness of the throat and dysphagia, together with a hoarse and stridulous cough, and aphonia. Pressure over the larynx gave pain. The pharynx had no abnormal appearance beyond redness, but the laryngoscope revealed deposits of gray false membrane on the surface of the red, swollen arytenoids, and in the interior of the larynx, covering the vocal cords, and giving them a jagged outline (Fig. 94). Temperature normal; pulse 80. The local tenderness, and the pain caused by coughing were the only matters of complaint; otherwise the patient felt perfectly well. After using inhalations of steam and slaking lime, a piece of tough, leathery false membrane was expectorated, which the microscope showed to be composed of coagulated fibrine in whose meshes were held white and red blood corpuscles.

The patient was kept in a warm room, and steamed hourly, flakes of false membrane being loosened and expelled every little while. Fœtor of the breath and a taste compared to that of carrion, were remedied so far as possible by carbolized sprays and gargles of potassic chlorate. The laryngoscope showed a steady amelioration of the local condition, till on the fifth day the larynx looked fairly clear, only a few spots of exudation remaining (Fig. 93). The general health seemed excellent, there being no more depression than might easily be referred to loss of rest consequent on painful cough. This improvement was only temporary, however, for on the sixth day a trace of albumen appeared in the urine, while the membrane re-formed and lined the larynx, creeping around the edges of the epiglottis, which became swollen and red. Deglution and coughing were now most painful and the sputa were tinged with blood. The pain was relieved by a spray of liq. morph. sulph. and atropia. The air of the chamber was kept charged with vapor, and the larynx poulticed. Two pints of urine (specific gravity 1025) continued to be passed daily, though the percentage of albumen rapidly increased, and the microscope detected numerous casts, but no blood. Diuretics produced no special effect. The temperature varied from 99½° to 100½°, 101° being its highest register. At no time was there any dyspnœa or head symptoms. The local lesion remained throughout confined to the larynx and upper part of the trachea. Inspection of the pharynx alone could have afforded no information whatever.

Improvement began again about the end of the second week. The temperature fell to $98\frac{1}{2}°$, the albuminuria diminished, the throat became less painful, the exudation was easily loosened and expectorated, and the appetite gradually returned. The laryngoscope showed a corresponding change for the better in the larynx, which, though swollen, was now free from exudation.

The patient gained ground daily. By the end of the month all traces of nephritis had disappeared. Aphonia remained, the larynx appearing normal except that the arytenoids were still a little swollen, and the vocal cords red, pitted and scarred by the ulceration which they had undergone. The voice was regained gradually, but has never completely recovered its normal flexibility.

Early in May the patient began to complain of tingling and numbness in the extremities and a considerable loss of power in the limbs, especially the lower. This paræsthesia and paresis increased till it became impossible for the patient to feel the reins when driving, or any object for which he might search his pockets; he had great difficulty in mounting stairs, and had to be assisted at his toilet. Yet he appeared in rugged health, and his weight had increased. These troubles have now left him, and the only traces of his disease discoverable are a chronic laryngitis, a few depressed scars on the vocal cords (Fig. 96), and a little huskiness of the voice, observed only occasionally. None of the attendants have suffered by contagion.

This case seems to me interesting because of the site at which the disease found local expression. It was throughout limited to the larynx and trachea, an occurrence certainly very uncommon. Moreover, its close clinical resemblance to what is termed membranous croup made the question of diagnosis for some little time a difficult one, though quarantine measures were enforced as if it had been undoubtedly diphtheria. The occurrence of albuminuria on the sixth day settled the question in my mind, and the sequelæ have made assurance doubly sure. Yet had the patient happened to die of asphyxia before the sixth day, I know of no feature of his disease which could have been cited as differentiating it from the membranous croup of childhood. There was no special depression, no swelling of cervical glands, no spreading by contagion. It seemed a purely local affair. And I could not but to ask myself: Supposing this were a young child, rebellious to treatment, lacking strength and skill to clear the air passages of this exudative material whose accumulation must speedily clog his small larynx and trachea, could I diagnosticate this disease from membranous croup, so called, or hope for any result other than the one so terribly familiar? Remembering the early death of patients with true croup, perhaps before systemic poisoning, if there were any, could indicate itself by albuminuria and how some cases of that disease seem to be acquired by one child from another through contagion it was a natural, though perhaps hasty outcome from the confusion of facts and theories, to adopt a conclusion of little theoretical importance, perhaps, but possessing considerable practical consequence, viz.: that membranous croup and laryngo-tracheal diphtheria may be indistinguishable, and that therefore all such cases should alike be quarantined.

A CASE OF TRAUMATIC AND SEPTIC EMPYEMA.

BY A. L. RANNEY, M.D.,

Adjunct Professor of Anatomy, University Medical College, New York.

The case represented by the accompanying plate came under my observation a few years since, and admirably represents the extreme type of ulceration which may follow suppuration within the cavity of the pleura. The patient was a girl who had received an injury directly applied to the chest, causing fracture of the fourth and fifth ribs of the left side, with pleuritic complications. She came under my care after being in the charge of other medical attendants for quite a lapse of time following the injury. During this period she had developed well-marked empyema, and the pus had appeared externally to the ribs, as fluctuating tumors which manifested a tendency to slough. Several spontaneous openings occurred rapidly, and through the larger ones the surface of the lung could be seen. The case was of such an extreme type that I made a sketch, which is reproduced in Fig. 97, as the best method of preserving a record of the situation and size of the openings. The drawing was made with every precaution against exaggeration; more pains being taken with the measurements than with the artistic portion of the work. The left side showed less retraction than would be supposed, with so extensive a destruction of the pleura. The patient gave evidence of the pyemic condition even before the spontaneous openings occurred, and eventually succumbed to that disease.

The following suggestions seem to me to be afforded by the facts related. The fracture of the ribs was evidently the exciting cause—yet such injuries do not generally result in so serious a complication. It may be well, therefore, to repeat a diagnostic point so often given, viz.: that fractures from *direct violence, applied at the seat of fracture*, are more liable to create pulmonary complications than those which follow compression of the chest or some other type of indirect violence. In the former the fragments are driven inward by the force of the blow; in the latter, the ribs bend before the fracture occurs, and the fragments tend to displace themselves outward.

Again, the existence of *septic poisoning* is one of the most frequent causes of extensive and fatal suppuration within the pleura. In this condition the proliferation of new cells is far more rapid than in normal inflammatory processes, and the pus tends to burrow rapidly, and to evacuate itself by spontaneous openings produced by ulceration of the soft tissues which enclose it.

[99]

In the third place, the percentage of recoveries in all forms of empyema is greater when the evacuation of pus is *spontaneously* accomplished than when incisions are resorted to for that purpose. The use of the aspirator for that matter, however, may often prevent the formation of an external opening, and thus possibly assist in causing recovery by the absorption of the fluid and the total arrest of the pus formation.

Fourthly, spontaneous evacuations of pus through the chest wall are not necessarily fatal, even when the openings are large and the lung exposed over a large area. Permanent fistulæ may form and pus escape in small quantities for years. The adjacent organs are liable to suffer marked displacement from their natural position, however, and the lung once subjected to external atmospheric pressure seldom regains its full expansibility under the most favorable auspices.

Fifthly, the evacuation of pus from the pleural cavity may not always be through the intercostal spaces. Cases are not infrequently reported where the lung itself has been made the vehicle for the spontaneous evacuation of pus. The diaphragm may also be perforated and the pus may then gravitate into the cavity of the abdomen and open spontaneously at a point far remote from the chest. These causes are often difficult of diagnosis. A fatal peritonitis may alone tell the medical attendant that pus has entered the peritoneal cavity; the escape of pus from the bowels reveals the fact that the alimentary canal has been made to communicate with the cavity of the pleura; a lumbar abscess shows that the abdominal wall has been the channel of purulent infiltration; a tumor in the groin points to the psoas muscle as the line of descent of fluid which has perforated the diaphragm.

Finally, the symptoms of a developing empyema are often extremely vague, and are liable to be overlooked until the disease has progressed far in its course. Local pain may be absent; the dyspnœa is usually slight; a previous pleurisy has yielded its characteristic symptoms which are not materially altered when the fluid in the chest begins to be transformed into pus; the change in the patient rather indicates that of the last stage of pulmonary phthisis; and at last the aspirator becomes indispensable in any endeavor to decide positively the question at issue.

[100]

TRAUMATIC AND SEPTIC EMPYEMA.
(Case of Prof. A. L. Rassett.)

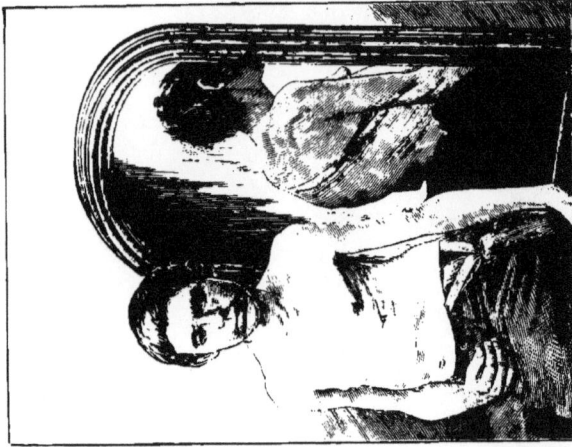

THORACIC DEFORMITY RESULTING FROM EMPYEMA.
(Case of J. E. Mears, M.D.)

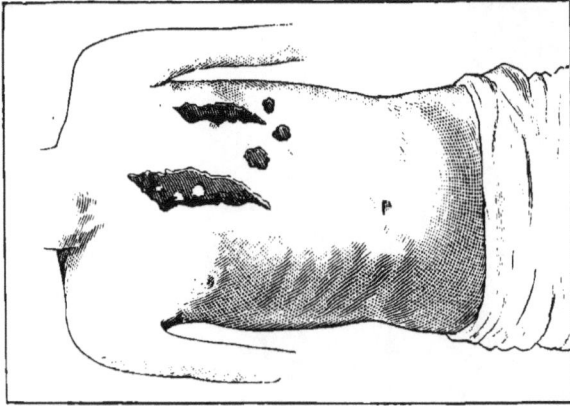

PLATE XVI.

REMARKABLE DEFORMITY FOLLOWING AN ATTACK OF EMPYEMA.

BY J. EWING MEARS, M.D.,

Surgeon to St. Mary's Hospital, Philadelphia, &c.

Thos. S., aet. 21, was admitted to the surgical wards of St. Mary's Hospital June 13, 1878. He had come to the city from the interior of the State for the purpose of receiving treatment for a condition of chronic empyema (which had followed an attack of acute pleurisy occurring two years previously), and in which the pus had escaped from the cavity of the pleura by a spontaneous opening. At the time of his admission he was in an extremely exhausted state and was very much emaciated. Quite a large quantity of pus was evacuated daily from the opening. On admission an examination was made, and the remarkable deformity was seen which is so well shown in Plate XVI. This distortion, he stated, had developed gradually, beginning a few weeks after his first illness; he suffered much pain upon the left side of the chest, which was relieved somewhat by inclining his body to that side.

Percussion revealed dullness over the entire surface of the left side of the chest, anteriorly and posteriorly. On auscultation a very indistict respiratory murmur could be detected at the apex of the left lung—at times it was scarcely audible.

The heart sounds were heard beneath the sternum and into the right side of the thorax; the cardiac action was very rapid, and it was thought that chronic pericarditis with adhesions existed.

Examination of the right lung showed that it was not involved, and that it was doing double duty. The thoracic wall on the left side was immovable during the acts of respiration.

The opening in the chest-wall, which had occurred spontaneously, was nearly on a line with the left nipple and between the sixth and seventh ribs, apparently—an unusual position for such openings to occur.

The treatment consisted in the administration of tonics, stimulants and nutritious food, with the injection into the thoracic cavity of carbolized water.

Syrup of the perchloride of iron was given in thirty-drop doses, with an ounce of cod-liver oil three times daily. Stimulants in the form of milk-punch were also given, with beef-tea and nutritious soups.

[101]

The opening in the chest-wall was enlarged by an incision, and a large drainage tube introduced and secured in position. Through the tube a five per cent. carbolized solution was injected twice daily, warm water being first thrown into the cavity to wash it out.

Under the treatment adopted the patient showed marked signs of improvement, his general health being improved and the discharge of pus diminishing in quantity. The treatment was continued for a period of three months, at the end of which time the discharge from the cavity was very slight in quantity, and it was thought that he could return to his home in the country with benefit as to the change in air. He was instructed to continue the injections and constitutional treatment as long as was deemed necessary by his attending physician. No effort was made to treat the deformity by the application of apparatus. No report has been received from him since his departure.

A CASE OF EXTRA-ARTICULAR TUMORS OF THE HANDS.

BY OLIVER P. REX, M.D.,

Visiting Physician to Jefferson Medical College Hospital, Philadelphia.

John B., æt. 42, born in England ; by occupation, a clerk ; history taken at Jefferson Medical College Hospital, May 26, 1882.

Nine years ago the joints of his right hand began to swell without pain. He attributed this to constant writing. He gave up writing for about one year, when his hand got better, but there was some deformity of the joints at the metacarpo-phalangeal articulation, the greatest deformity being at the junction of the index finger with the metacarpal bone. His family were all very healthy, with no history of gout. About the 1st of May, 1882, he left England for America on board an emigrant ship, and, while at sea, had but little food, and that of the poorest kind. He had not been on this diet long, when the joints of both hands began to swell "without pain." A tumor also appeared on the right hand, about the middle of the second metacarpal bone. When it made its first appearance, it was very small, but grew rapidly from day to day.

Examination showed this tumor to be about the size of an almond, on dorsal surface of hand, movable, and not attached to surrounding parts. On palpation it was found to be multilocular, and, when firmly pressed, gave the peculiar sensation of containing sand. The right ring finger, at the second phalangeal articulation, was very much swollen, red, and slightly painful. On careful examination it was found that the swelling was extra-articular, not involving the joint nor interfering with its function. The index finger of same hand was involved at the metacarpo-phalangeal

Fig. 99.

articulation, and also at the phalangeal articulation of second joint, both being extra-articular. The left hand was also involved at the base of index finger and junction of metacarpal bone. This was the only joint that had the appearance of being involved, and its function interfered with. The little finger was also very large and swollen at the second phalangeal articulation, very red, with an abrasion at its apex, exuding a thin, transparent fluid. There was no enlargement on the dorsal

[103]

surface of left hand. Patient's general appearance indicated that he had been poorly nourished. He was put upon good, nourishing food, and carbonate of lithia, gr. ii., t. i. d., was prescribed. Under this treatment the patient improved rapidly in strength, and he was apparently in good health at the end of four weeks; but the enlargement of the joints and tumors on hand did not diminish in the least. Surgical interference was now deemed proper. In the meantime, Dr. Longstreth had examined fluid taken from the tumor on the dorsal surface of hand by an exploring needle, and found a large amount of the crystals of nitrate of soda. On the first of July *Dr. Lewis* made a free incision in the tumor on the dorsum of right hand, and allowed a free discharge of a white chalky substance. On the second day after the operation, the tissues surrounding the wound becoming red and swollen, with some pain, lead water and laudanum were applied. The inflammation soon subsided, and the wound healed rapidly. The enlargements on the fingers were treated in the same manner, with good result, with the exception of the enlargement on little finger of the left hand. Pure carbolic acid was injected into this without good result, free incision being the only treatment that was attended with success. On the 22d of July the patient was discharged, to all appearance cured.

[104]

VOLUME II. No. 1. 1883.

ILLUSTRATED

MEDICINE AND SURGERY.

EDITED BY

GEORGE HENRY FOX,
CLINICAL PROFESSOR OF DISEASES OF THE SKIN,
COLLEGE OF PHYSICIANS AND SURGEONS, NEW YORK.

AND

FREDERIC R. STURGIS,
PROFESSOR OF VENEREAL AND GENITO-URINARY DISEASES,
POST GRADUATE MEDICAL SCHOOL OF NEW YORK.

WITH THE CO-OPERATION OF

PROFESSORS WILLARD PARKER, A. C. POST, W. H. VAN BUREN, J. L. LITTLE, T. G. THOMAS,
A. L. LOOMIS, F. DELAFIELD, D. B. ST. J. ROOSA,
C. R. AGNEW, AND AUSTIN FLINT.

CONTENTS.

QUARTERLY.

NEW YORK:

E. B. TREAT, No. 757 Broadway,

TRUBNER & CO., LONDON,
57 & 59 LUDGATE HILL.

[COPYRIGHT. 1883.]

J. B. BAILLIÈRE & SONS, PARIS,
19 RUE HAUTFFEUILLE

Entered at the Post Office, New York, as Second-Class Matter.

ILLUSTRATED

MEDICINE&SURGERY

1884.

CONTENTS.—VOL. II.

CONTENTS.—VOL. II.

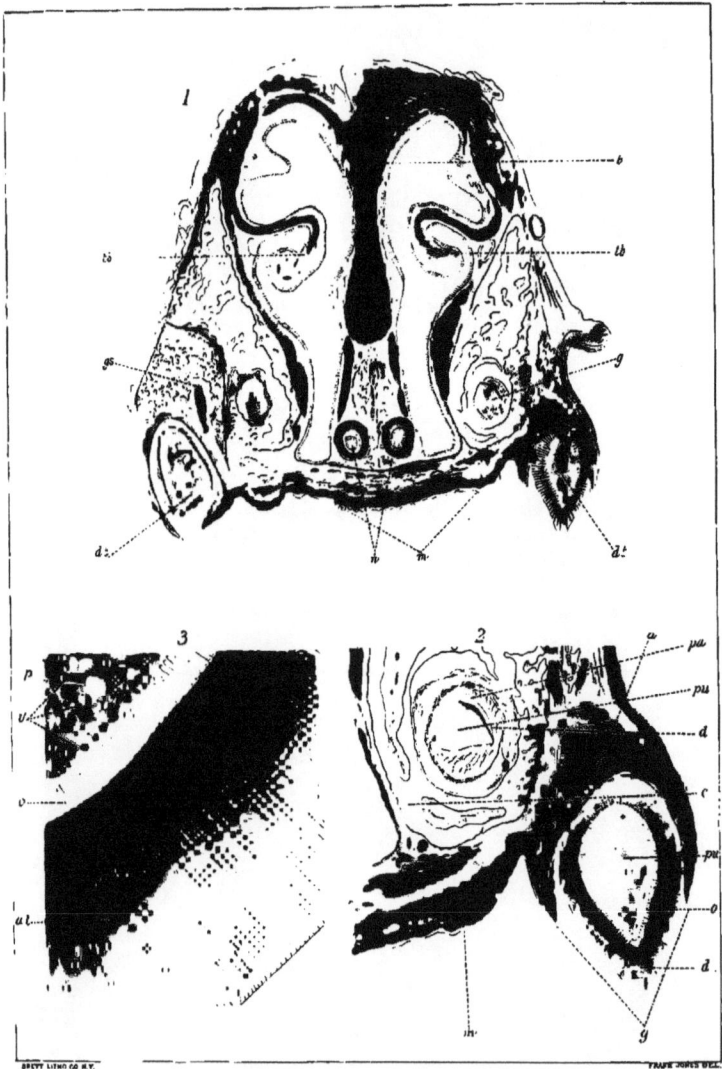

XVII DENTAL DEVELOPMENT.
Microscopic sections.
(By Prof. W.m Hailes Jr.)

DENTAL DEVELOPMENT.

BY WM. HAILES, JR., M.D.,

Prof. of Histology and Pathological Anatomy, Albany Medical College.

The following contribution, together with the accompanying illustration, is one of a contemplated series of articles which, through the courtesy of the editors, will appear from time to time in their valuable journal:

The subjects will be in normal and pathological histology and embryology, the object being to call attention to methods available for instructing classes in practical laboratory work, demonstrating the feasibility of making comprehensive sections which not only give the topography of the parts and their important relations to each other, but also permit a most satisfactory minute examination in any part of the section, having the uniform thickness of about $\frac{1}{1200}$ of an inch.

We have chosen for our illustration a section through the superior maxilla of a kitten two weeks old.

The jaws were decalcified by means of picric acid, were placed in mucilage, and cut in the freezing microtome into hundreds of sections, each about $\frac{1}{1200}$ of an inch in thickness. These sections are kept preserved in alcohol for future use in the laboratory.

Any section taken at random from the bottle, will illustrate perfectly the points necessary to be seen in studying the structure of growing bone, teeth, etc., etc., differing slightly, however, according to the locality from which the section was taken, whether canine, premolar, or molar.

The drawing accompanying this article is from a section directly through the canine teeth, and exhibits beautifully not only the general structure and arrangement of the jaws—teeth (both temporary and permanent)—but also the nasal passages, vomer, turbinated bones, mucous membrane, etc., besides demonstrating perfectly the minute structure of pulp-odontoblasts and dentine, and exhibits the germ of the permanent canine inclosed in its special osseous chamber.

The sac of the permanent tooth, after a certain stage of its development, adheres to the back of the sac of the temporary tooth. " Both of them continue to grow rapidly, and after a time it is found that the bony socket not only forms a cell for the reception of the milk-sac, but also a small posterior recess or niche for the permanent tooth-sac, with which the recess keeps pace in its growth; it is found that at length the permanent saw so far recedes in the bone as to be lodged in a special osseous chamber, at some distance below and behind the milk-teeth, the two being completely separated from each other by a bony partition." *   (See Fig. 101.)

Many other subjects of greater interest might have been chosen from sections now in the laboratory, for after ten years or more of work in the teaching of practical microscopy in the laboratory of a medical school, one learns to simplify methods and make improvements, and to bring within the range of accomplishment many things hitherto beyond the reach of the student, because of the time required for their preparation and study.

In this work I have been greatly assisted by my friend and assistant, Dr. S. G. Shanks, of Albany. I am also greatly indebted to my friend, C. E. Beecher, of N. Y. State Museum, Albany, in the preparation of the drawings.

There are many other points of interest in this section, which an inspection of the Figs. will exhibit.

Explanation of Figures.—Superior Maxilla of Kitten. Fig. 100. *v*, vomer; *t b*, turbinated bone; *d t*, deciduous or milk teeth; *g*, germ of permanent canine; *n*, nasal passages, lined by mucous membrane and covered by ciliated epithelium. -- X 9.

Fig. 101. Canine (temporary and permanent). *pn*, pulp; *o*, odontoblasts; *d*, dentine; *c*, osseous chamber for permanent tooth-germ; *m*, mucous membrane of roof of mouth; *c*, gum.—X 18.

Fig. 102. Canine, x 200. *p*, pulp; *v*, vessels; *o*, odontoblasts, showing processes, etc.; *d t*, dental tubules.

* Quain's Anatomy.

A CASE OF PALATO-PHARYNGEAL SARCOMA.

BY JOHNSON ELIOT, M.D.,

Emeritus Professor of Surgery, Medical Department of Georgetown College. Surgeon to Providence Hospital, Etc., Washington, D. C.

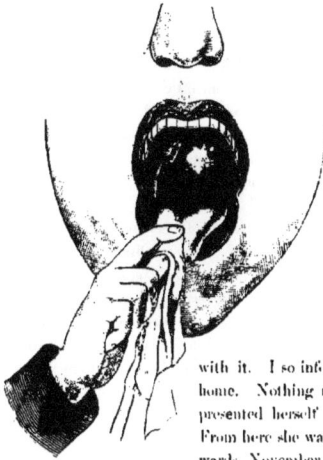

In the summer of 1880, Miss F. V. White, aged twenty-three, was brought to Providence Hospital for examination. Her general appearance indicated impaired health. She was anemic, feeble and emaciated. Examination revealed a tumor the size of an English walnut, situated upon the left side of the palate, extending to the pharynx and tonsil of the same side. She informed me that it had been growing for more than a year. Her voice was husky, deglutition painful, and respiration slightly embarrassed. The tumor was elastic on pressure, painful to the touch, and several deeply ulcerated points were noticed on its surface. The lymphatic glands of the corresponding side were enlarged and sensitive. The tumor was diagnosticated as malignant, and it was not deemed advisable to interfere with it. I so informed her mother, and advised that she be taken to her home. Nothing more was heard of her until November, 1881, when she presented herself at the Central Dispensary in this city for treatment. From here she was sent again to Providence Hospital. A short time afterwards, November 1st, she was admitted, with the determination to submit to any treatment that might offer even temporary relief. Her general condition was somewhat improved since her first visit to the hospital, but the local trouble had greatly increased. The tumor had increased in size; its pressure upon the palate and surrounding parts rendered her respiration, particularly in the recumbent position, painful and embarrassed; it was impossible to close the mouth, and there was great salivation. The ulcerations had cicatrized; the surface had become smooth; the lymphatic glands of the neck had decreased in size. With the local changes for the better in the tumor and her generally improved condition, we concluded that the growth was not as malignant as I at first supposed. Under the existing circumstances it was deemed advisable to operate.

On November 22d, 1881, in the presence of Drs. Ashford, Magruder, Carroll, Morgan, Bayne and Mallan, I proceeded to operate. The patient was seated on a chair, with her arms pinioned. Anæsthetics were not used for evident reasons. From the extreme vascularity of the region and the seemingly congested appearance of the tumor, hæmorrhage was anticipated, and ample precautions were taken for its suppression in the event of its occurring. Hæmostatics were at hand, the thermo-cautery was within reach.

An incision was made in the long diameter of the tumor, and its contents were rapidly enucleated. Several small pieces of bone came away with it. To our surprise, but little hæmorrhage followed; no ligatures were required, and no hæmostatics were applied. The bleeding ceased with the conclusion of the operation. On the second day after the operation hæmorrhage occurred, profuse but passive in character. Dr. Mallan, house surgeon, plugged the cavity with lint saturated with Monsel's solution of iron, and the hæmorrhage was finally controlled. The patient convalesced slowly. After several weeks in the hospital she left.

February 16th, 1882. Greatly improved in health, and relieved of all her former suffering. I received several letters from her while at home, writing encouragingly of her condition, and congratulating herself on her returning health. To my surprise she returned to the hospital on March 12th, 1882. An examination revealed a small swelling at the side of the cicatrix; it gave her but little inconvenience; she was concerned about it, however, and was anxious to have it removed. I intended to remove it at once. On the day fixed for the operation, March 13th, while conversing with a patient in the ward, without premonition violent hæmorrhage took place, and death followed immediately. No autopsy was made. The presumption is that some large vessel was opened by the softening of the tissue and yielded to the arterial pressure. Microscopical examination demonstrated the tumor to be a spindle-cell sarcoma.

EXCISION OF THE SHOULDER JOINT.

BY RANDOLPH WINSLOW, M.D.,

Demonstrator of Anatomy in the University of Maryland, and Professor of Surgery in the
Women's Medical College of Baltimore.

CHARLES C. SMITH, negro, aged 39 years, a rag-picker by trade, was admitted to the University Hospital on May 29th, 1882, with the diagnosis of caries of the shoulder joint.

His sickness dates back about ten months, and is supposed to be consequent to rheumatic arthritis.

Condition upon admission: Patient somewhat emaciated, and exhibiting symptoms of hectic. Has lost all use of his left arm, and as he walks he supports the diseased arm with the sound hand. Has pain, worse at night and upon any motion of the part. The left shoulder is flatter than normal, and the caput humeri is partially dislocated under the coracoid process. Several sinuses are found upon the posterior aspect of the arm, just behind the insertion of the deltoid muscle, and two others are present in the axilla. From these openings pus constantly exudes.

May 31st. The patient was anaesthetized, and probes were introduced into the sinuses in the expectation of finding carious bone, but as no probe in my possession was long enough to reach the diseased surface, I contented myself with establishing better drainage by straightening some of the sinuses, and inserting tubes in various directions.

June 14. Caries of the head of the humerus was detected by a long probe, and excision was determined upon.

June 16. The head and one inch of the shaft of the humerus was excised by a single straight incision, extending from the anterior border of the acromion process almost to the insertion of the deltoid, and passing through the anterior portion of this muscle. The periosteum and capsular ligament were divided longitudinally, and with the insertions of the pericapsular muscles were separated

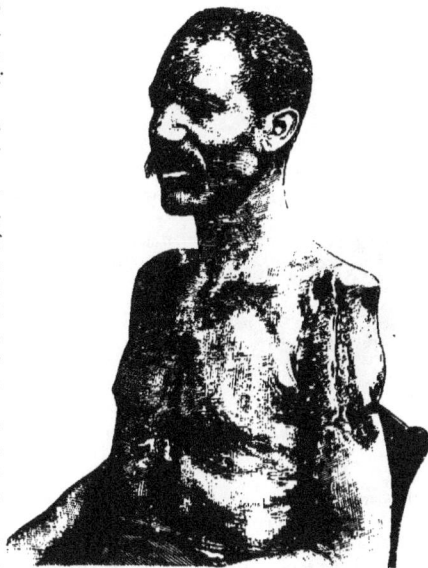

[109]

from the bone by periosteal elevator and scalpel, and retained. The detachment of the periosteum was a difficult process on account of the large amount of osteophytic growth which was present. The long head of the biceps had disappeared. The glenoid cavity had become obliterated and was filled up by a granulating mass, and no caries of the portion of the joint could be detected. The head of the humerus was protruded with some difficulty, and the section was made with a chain saw about an inch below the surgical neck. The specimen removed was carious at its posterior aspect, where there was a small undetached sequestrum, and anteriorly there were deep excavations behind the tuberosities.

The wound was thoroughly irrigated with a carbolized lotion, and a drainage tube passed entirely through the cavity and brought out posteriorly through one of the previously-existing sinuses. The wound was closed with wire sutures and dressed externally with iodoform, a pad of carbolized oakum surrounding the whole. The arm was extended upon a board splint at right angles from the body for the first few days; subsequently it was simply supported in a sling under the elbow and fore-arm.

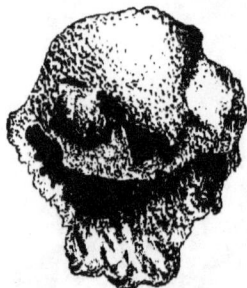

June 20. All the sutures removed, and union by the first intention was found to have taken place except at the point of entrance of the drainage tube at the upper portion of the wound. The large cavity left after the removal of the bone was kept clean and disinfected by carbolized lotions injected through the tube. The highest temperature reached was 101½ on the fourth day, which fell to 99½ on the fifth day, and never exceeded 100¾ subsequently.

Fig. 105.

Fig. 106.

After the line of incision had healed several sinuses formed in the cicatrix, probably due to the retention of some discharge after the removal of the drainage tube. These tracks were laid open and healed from the bottom, except one through which a small quantity of serum still escapes. All the pre-existing openings have gradually closed.

Result of the operation: The patient's general health is restored; he eats heartily, sleeps well, and is in good flesh. The functions of the forearm are perfect, and its muscles well developed. Motion between humerus and scapula limited; can abduct the elbow from the side nine inches; has considerable backward motion, but has not much power to push the arm forward beyond the plane of the body.

The man is able to lift heavy weights and to pursue his business. He has a useful member but an impaired one. He was rapidly running down and could not have waited for a cure by natural processes, even if that had been possible, and the choice lay between amputation at the shoulder or excision of that joint. I chose the latter, and the result justifies the procedure.

THREE CASES OF COMPOUND COMPLICATED HARE-LIP,

OCCURRING IN THE SAME FAMILY; OPERATIONS FOR RESTORATION OF LIPS; WITH
REMARKS ON THE OPERATION FOR CLEFT PALATE.

BY JAMES L. LITTLE, M.D.,

*Professor of Clinical and Operative Surgery in the New York Post-Graduate Medical School ; Professor
of Surgery in the Medical Department of the University of Vermont : Surgeon to St.
Luke's and St. Vincent's Hospitals, New York City, &c., &c.*

The cases will be described in the order in which they came under my observation:

WILLIAM Bocock, aged 21.　　JOHN Bocock, aged 9.　　CHARLES Bocock, aged 18.

No hereditary tendency can be traced in father or mother's family. There were four boys and five girls. All the boys were born with hare-lip, while no deformity existed in any of the girls.

The order in which the children were born is as follows:

1. William. Compound complicated hare-lip.　2. A girl with no deformity.

3. Charles. Compound complicated hare-lip. A spindle-celled sarcoma made its appearance on the left side of the perineum in 1878, which I removed. It recurred, and I again removed it in October, 1882.　4. Girl with no deformity.　5. Girl with no deformity.

6. John. Compound complicated hare-lip; absence of ring-finger of right hand.

7. Girl with no deformity.　8. Girl with no deformity.

9. Boy with single hare-lip, who died in infancy.

These patients presented this deformity in almost the worst possible form, the arrest of development occurring at a very early period of fœtal life. The inter-maxillary bone in each case was distinct, being ununited to the superior maxillaries, and was continuous with the nasal septum and vomer. This projecting bone was partially covered by a tag of integument, which was continuous with that of the tip of the nose. In John (Case II), this bone contained two well-developed incisor teeth, while in William and Charles (Cases I and III), there was but one. There was a complete absence of both the hard and soft palate in all the cases, and in Case III the fissure was unusually wide (4 *cm.*). Articulation was so imperfect that they could be understood with the greatest difficulty.

Operations.—Case I.—William. Uranoplasty was performed at St. Luke's Hospital on February 9th, 1878. The sides of the fissure in the hard palate being imperfectly developed, and running obliquely upwards and inwards, the only operation that could be performed was the dissecting of the soft parts from the bone, from above downwards, and allowing them to meet at *B*, Fig. 113,

This was done, and there was sufficient material for the flaps to overlap in the central line. A small strip was removed from each flap and the edges accurately joined together. The union was complete throughout. The sutures were removed, and I felt sure of a satisfactory result. In about a week I found the line of union growing thin and showing indications of breaking away. An examination with the laryngeal mirror in the nasal cavity showed that granulations were springing up at the angles *e e*, and a line of ulceration was visible along the nasal surface of the wound, *B*. The union of the flaps finally gave way, and they assumed almost their original position in contact with the bony walls, *A A*, with the exception of a piece about one inch in width, at a point near the posterior edge of the hard palate. No further operation was permitted for the relief of this part of the deformity.

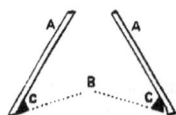

Fig. 113.

Operation for the Restoration of the Lip.—March 21st, 1878. The piece of integument covering the inter-maxillary bone was dissected up to the tip of the nose, and the projecting bone containing the incisor tooth was removed with bone forceps. The piece of integument which had been lifted up was then trimmed, turned down and accurately adjusted against the raw edge of the septum, so as to form a columna; this was retained in place by two silver wire and shot sutures, running directly through the septum. The union was perfect. On April 17th the following operation was performed:

The two portions of the lip were separated freely from the superior maxillary bones and a piece was removed on either side of the cleft by a pair of curved scissors. The posterior extremity of the newly formed columna was freshened, the edges of the fissure were brought together and retained in position by two pin sutures and a number of fine interrupted silk sutures. The columna was also nicely adjusted in the upper portion of the fissure. Good union took place everywhere except at the columna and the right side of the lip, and a considerable notch was also left in the centre of the lip. The boy went home during the summer to allow the parts to become soft and more yielding. He returned to the hospital December 9th, 1878. The edges of the opening between the columna and lip were freshened and closed with sutures. The union was perfect. On December 27th the final operation was performed on this case, and was for the purpose of relieving the notch at the line of union of the lip. Wharton Jones' operation, consisting of two incisions through the entire thickness of lip, on either side of the median line, extending from a point just below the columna downwards and outwards to but not through the vermilion border, forming an inverted V incision, was performed. The V-shaped flap included between these incisions was then pulled down, so that the notch was obliterated; the upper portion of the incision was closed by a pin suture, and the remaining portions by fine silk sutures. The lines of the incisions left after this operation was of the form of an inverted Y. The final result is illustrated in Plate XVIII, Case I.

Case II.—John, aged 9. The operation performed December 12, 1878, for the formation of a new columna, was the same as in Case No. I. On January 10th, 1879, the columna formed by the first operation being too prominent, its integument was dissected up and replaced after another portion of the bony septum had been removed, resulting in a well-formed columna. The operation on the lip was the same as in Case I, resulting in perfect union. Its appearance is shown in Plate XVIII, Case II.

difference was scarcely appreciable. This boy also had a very peculiar formation of the right hand, consisting in the absence of the ring-finger and its metacarpal bone. The little finger sprang off at right angles to the hand, just below the line of the carpal bones, and I cannot say whether it had a short metacarpal bone or not. It could be flexed in the manner shown in Fig. 114.

Case III.—Charles, aged 18. In this case the inter-maxillary bone was in contact with, although not united to, the right maxilla. The orifice of the right nostril was perfectly formed; the fissure in the lip and hard palate was unusually large, measuring four cm. in width. The mucous membrane covering the left inferior turbinated bone was hypertrophied, forming a prominent mass in the nostril. The operation was performed as in the preceding cases, with this exception: the columna being formed and the lips brought together at one operation. The union was perfect.

A supplementary operation was afterward performed. The left ala nasi being too widely separated from the septum, a V-shaped piece was removed be-

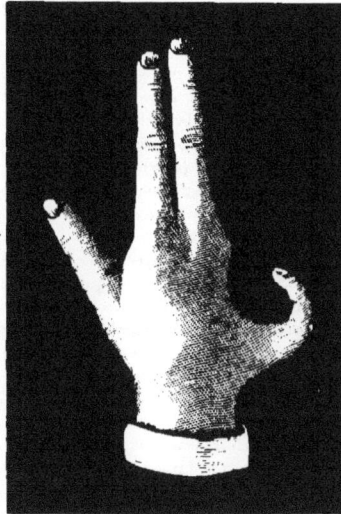

FIG. 114.

tween these two points, which were then brought together so as to form a perfectly oval nostril. A piece was also removed from the right side of the columna, reducing its thickness. The patient was discharged in the condition seen in Plate XVIII, Case III, still having a slight notch on the margin of the lip. Three years after, I found that all traces of the notch had disappeared.

In concluding this paper I desire to say a few words regarding Uranoplasty and Staphylorraphy. I had performed these operations successfully a number of times before operating upon the case described in the first part of this paper. Since that time I have carefully looked into the results, and find that although in a large proportion of the cases the operations are successful so far as the closure of the fissure in the hard and soft palate is concerned, yet so little, if any, benefit is obtained in the improvement of the articulation, that I have been forced to the conclusion that they should be discarded as surgical procedures in adults. I refer of course to cases in which the cleft is congenital. Mr. George Pollock says "the real object of the operation of closing the cleft in the palate is to enable the patient to articulate hereafter, plainly and intelligibly—*not* to enable the child to take food." * This last difficulty, he

* Holmes' Surgery, 2d Edition, V. I, IV, p. 429.

[113]

states, is overcome in a few days. My three patients had no trouble whatever in this respect. What the result would be if the operation was performed in early life, I have no means of knowing. From my own experience in operations upon the adult I can conclusively state that no improvement has ever taken place in the patient's articulation. The reason is undoubtedly this: the newly formed palate is rigid, tense, and deficient in length, and in a large majority of cases it cannot be brought into apposition to the pharyngeal wall, so as to close the buccal from the nasal cavities; and unless this be done perfect articulation becomes impossible. The division of the palatine muscles, which is necessary in the performance of this operation, also interferes, to a certain extent, with the proper use of the organs in speech.

While I have never seen a case in which the nasal twang was improved, I have seen a number of patients in whom an artificial palate rendered the articulation absolutely perfect. Norman W. Kingsley, M.D.S., D.D.S., of this city, who has paid a great deal of attention to this subject, has invented a soft artificial velum, which is so under the control of the surrounding and adjacent muscles, opening and closing the passages at will, that the wearer, after a certain amount of practice and education, acquires a perfect articulation. In one of my cases (No. III), Dr. Kingsley applied one of his artificial palates. Before introducing it an experiment was performed in my office, which I will describe in Dr. Kingsley's own words:

" Altogether the most extensive deformity of this kind, and the one having the most disastrous influence on the speech that I have ever seen, was a young man upon whom Professor J. L. Little, of the College of Physicians and Surgeons, New York, operated for compound hare-lip, and who afterwards came into my hands for an artificial palate. * * * Previous to the introduction of the artificial palate the following experiment was tried in the presence of a number of well-known surgeons: I wrote upon a slip of paper the following syllables, which the patient pronounced to the best of his ability, repeating each one several times: Bo, Lo, Ho, Mo, Ko, Po, Go, No, &c. The sound given by him to each of these syllables was written by the gentlemen present as nearly as they could be understood. A comparison of the various records showed that the only unmistakable syllables of the whole list were Ko, Go, and Ho, all throat sounds. Of the doubtful ones, No and Mo were interchangeable, and so were Lo and Ko; and of all the others no sound that he gave was any clew to the syllable he was trying to pronounce." *

The velum was applied, and the patient remained under Dr. Kingsley's instruction about two weeks only. He then returned to his home, where he was practiced in correct pronunciation by Dr. Levi W. Case. He returned, in three months, and was able to read a page from a medical journal before a number of physicians so that every word was understood.

I would urge, however, that the operation be performed if possible in early life, as recommended by Mr. Thomas Smith of St. Bartholomew's Hospital,† with the hope that, as the organs develop with the growth of the child, this difficulty will be overcome and artificial means dispensed with.

* A Treatise on Oral Deformities, p. 402.   † Transactions Medico-Chirurgical Society, Vol. LI., p. 79.

CYSTO-ADENOMA OF THE THYROID GLAND.

BY CHARLES BUCKLEY, M.D.,

Member of the Rochester Pathological Society.

The accompanying plates, Figs. 115, 116, represent the following case

In April, 1882, I was called to see Miss H. S., aged 56 years, a native of Germany, who complained of difficult breathing, incessant cough, inability to rest in the recumbent position, severe attacks of asthma, and painful deglutition. The patient gave the following history: Ten years ago she observed a swelling the size of a hickory-nut, situated in the left lobe of the thyroid gland. During the succeeding years it continued to enlarge, notwithstanding energetic treatment.

Fig. 115.   Fig. 116.

On examination a large growth was found occupying the front and sides of the neck, extending from the anterior edge of the right sterno-cleido-mastoid muscle across the larynx to the superior carotid triangle of the left side, displacing the carotid artery, jugular vein, and sterno-cleido-mastoid muscle backward. The carotid artery was found in a position corresponding to a vertical line drawn from the

[115]

CYSTO-ADENOMA OF THE THYROID GLAND.

lobe of the left ear. Its inferior border projected below the clavicle externally, and internally it occupied the subclavian triangle, resting on the apex of the left pleura. The integument was movable and contained a number of dilated veins.

On physical examination moist râles were found in the bronchi of both lungs, with dullness on percussion at the apex of left lung. The difficult respiration was gradually increasing, and she was slowly but surely becoming suffocated. At her earnest solicitation I consented to attempt the removal of the mass. On May 20th, 1882, the operation was commenced by making an incision six inches in length through the integument and carefully dissecting down. The left posterior border of the tumor rested against the cervical vertebra; the posterior border pressed against the larynx, to which it was firmly attached. It was freely supplied by the superior thyroid artery, and by the ascending cervical and transversalis humeri arteries, which were ligated in close proximity to the axis, pains being taken not to injure the recurrent laryngeal nerve. The operation consumed 2½ hours. The pressure caused by the growth being removed from the apex of the lung, the patient was seized with a violent paroxysm of coughing, which caused a rupture of the pleura, and air exuded at each expiration through the aperture of the pleura and wound. The wound was carefully dressed antiseptically, the edges were brought in apposition by sutures, and a compress applied. Toward evening the patient was attacked with asthma, which caused severe venous hemorrhage. The wound was partially reopened to secure the bleeding vessels, when the apex of the lung was observed at each expiration to ascend above the clavicle about 1½ inches; during inspiration the lung descended into the thoracic cavity. The patient made a good recovery.

It is the opinion of the writer that, notwithstanding authorities condemn interference with tumors involving the thyroid gland, an operative procedure was justifiable in this case.

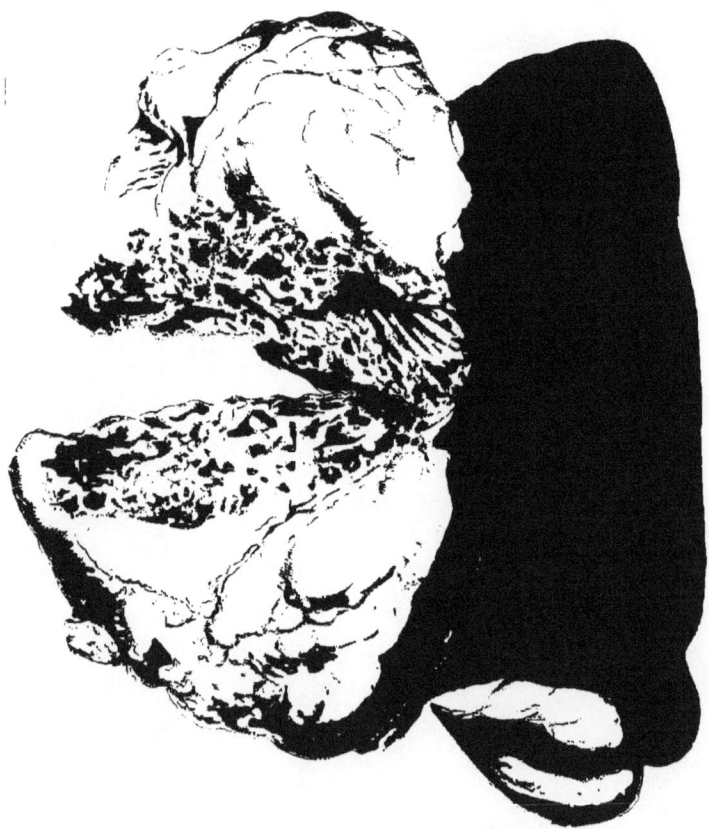

SECONDARY MYELOID DISEASE OF PLEURA AND LUNG.

BY WILLIAM OSLER, M.D., M.R.C.P.,

Professor of the Institute of Medicine, McGill University; Pathologist to the General Hospital, Montreal.

The following case is of interest from the extensive secondary affection of the lung and pleura. My colleague, Dr. George Ross, has kindly furnished the clinical report :

B. M., æt. 15, came under my care during the month of October, 1880. He was then complaining of shortness of breath, and had a rather troublesome cough. In the month of March previously he had been admitted into the General Hospital, under Dr. Roddick, for a large swelling at the left knee. This was diagnosticated as myeloid tumor, and the limb was amputated in the middle of the thigh. Subsequent examination proved that such was the nature of the growth. Whilst he had been under the care of the surgeons he had not complained of any chest symptoms, and the physical examination of these regions was entirely negative.

On admission it was found that the degree of dyspnœa was considerable. He was propped up with pillows and became distressed if his head was lowered. His cough was particularly annoying at night and prevented him from sleeping. There was a small quantity of muco-purulent expectoration. He suffered no pain.

The chest seemed wanting in expansive movement. The percussion was absolutely flat from the left clavicle downwards till past the cardiac area, and laterally to the edge of the axilla. The sternal and right infra-clavicular regions were also flat. No chest-sounds were to be heard over this area, and the vocal fremitus was absent. The heart-sounds were also much obscured. The diagnosis was a secondary tumor within the thorax, and it was thought probable that some pericardial effusion co-existed with it.

From this time the dyspnœa became progressively greater until it reached a most intense degree, and severe suffocative attacks ensued, and in one of these he died in January, 1881. No other symptoms or signs of intrathoracic pressure were observed, with one special exception, viz., enlargement of the veins of the front of the thorax. When first seen these were faintly traceable under the skin. They soon began to dilate, and for some time before his death some of them stood out as great tortuous, varicose vessels, as large as the little finger.

[117]

SECONDARY MYELOID DISEASE OF PLEURA AND LUNG.

Autopsy.—Body emaciated. No return of disease in the stump. Morbid condition confined to the chest. On removal of the sternum the mass, represented in the plate, was seen projecting from the anterior border and under surface of the upper lobe of the left lung, covering the pericardium, pushing back and compressing the lung, and occupying the greater part of the mammary region. It was not closely adherent to the sternum or ribs in front; behind it was attached to and grew from the pleura, and had extended for about an inch into the lower part of the anterior border of the upper lobe. Anterior surface was irregular, presenting grayish-white, fleshy prominences, mixed with areas of a dark maroon color. On section the tumor presented an alveolated appearance from the number of irregular cysts scattered through it. They varied in size from a pea to a marble, and contained a reddish fluid. In the centre of the growth the texture had a brownish color and was interspersed with hemorrhages. The upper lobe was compressed and airless, the lower crepitant; no secondary masses. Between the diaphragm and the pleura was a mass the size of an apple, having the same general characters as the larger one.

Right Lung.—A large flattened mass covered the upper lobes in front and above, and was closely united to the costal pleura. The lower lobe was free, but behind, near the spine, there was a rounded growth the size of an orange. On removal, the upper half of the organ looked rough and shreddy where it had been torn away from the ribs, the impressions of which crossed it transversely. The tissue was reddish-brown, with white, fleshy-looking masses scattered through it, and here and there a dark sanguineous cyst. Section of the upper lobe showed that its tissue was invaded; the hinder portion alone remained and was flattened. Upon the middle lobe were two firm, grayish-white masses, which on section were like recently formed spongy bone, reddish in color and with numerous ossific spicules scattered through the tissue. Two smaller growths of a similar character were on the anterior margin of lower lobe. The rounded tumor near the spine presented osseous spots at the periphery, while the central part was occupied by a large hemorrhage. The bronchial glands were not involved, nor were the lymph glands in the groins or abdomen.

The histological characters of the growths corresponded to myeloid sarcoma. The chief elements

were spindle cells, among which the multinuclear giant cells were imbedded. (Fig. 118.) In many regions these were scanty or absent, and the growth was like a spindle-celled sarcoma. This was most evident in the masses on and in the right lung. In the mass on the left pleura the condition on section, with the numerous cystic spaces, was just such as Mr. Gray describes as myelo-cystic tumors. The giant-celled sarcomata are not often followed by secondary growths; but when they do occur, the lungs are apt to be the seat of the recurrence. This case is similar, in many respects, to one published by Dr. Wilks. (*Path. Transactions*, Vol. IX.)

Fig. 118.

[118]

CONGENITAL UNION OF THE FINGERS.

WITH

TWO OTHER CASES OF MALFORMATION OF THE HAND.

BY J. H. POOLEY, M.D.,

Formerly Professor of Surgery in Starling Medical College. Professor of Medical Jurisprudence in the Columbus Medical College.

THOMAS E. B., aged sixteen, a native of the United States, came under my care May 20, 1874. He was the subject of a very interesting and complete congenital fusion of the phalanges of both hands and feet. Syndactilitis universalis. As one hand and foot was an almost exact duplicate of the other, a description of one will suffice for an understanding of both.

Taking, then, the right hand for description, we find the following peculiarities: *Dorsal aspect*—1st. The ring and middle fingers are entirely united from their commissure to their extremities.

While normally the middle finger projects somewhat beyond the index and ring fingers, in this case it is not so, as the extremities of the united fingers are on a level with each other, being made so by their close union, and yet the middle finger is of its normal length. It is therefore curved, with its convexity outward, on the back of the hand, where it rises or projects above its neighbors; and it is plain that if the unnatural connection were severed it would project beyond the digit on either side of it, as in a normal hand.

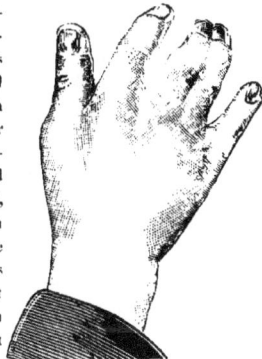

Fig. 119.

Fig. 120.

[119]

The line of union is even with the dorsum of the hand, except where the knuckles of the middle finger rises above its neighbor. The thickness of the union is equal to the thickness of the fingers themselves; it is not a mere web. The two nails seem to arise from one matrix; they lie deeply imbedded in the skin; the line of separation between them is marked by a faint depression. All along the terminal or ungual phalanx the union is very firm, apparently bony.

2. The commissure between the index and middle finger is extended as far as the first phalangeal articulation, forming a long web between these two fingers. Originally they were joined together like

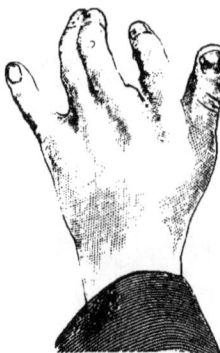

the two fingers already described, and the remaining web is the result of a partially successful operation performed in infancy; the little finger is normal in every respect.

Between the thumb and the index there is a thick web, extending as far as the middle of the first phalanx of the thumb. A cicatrix here shows the trace of a partial attempt to separate what was a complete union, performed at the same time as the operation already referred to, and even less successful in its results.

Fig. 121

Fig. 122.

The whole palmar surface presents a deep cup-shaped hollow, from the necessary curvature of the united fingers as already described.

The skin of the palm is thick and hard. The feet present the same kind of deformity as the hands, but not in so marked a degree; and as they do not interfere with locomotion, are of no practical interest or importance, and call for no treatment. The patient has no knowledge of similar, or indeed of any malformation in his family, on either side.

The illustration, from a photograph, represents the dorsal and palmar aspect of both hands, and gives a very fair idea of the condition of the patient. Considering the difficulty of keeping apart the surfaces of united digits after their separation, the previous unsuccessful attempts made in this case and the extensive nature of the attachment, I determined to proceed in a different way; to make a flap from the dorsal surface of one finger and the palmar surface of the other respectively, and fold them over, so that each would cover the raw surface of the other finger.

This operation is described in "Annandale on Malformation of the Fingers and Toes."

Erichsen barely alludes to it in his "System of Surgery," but gives a diagram of the operation, which is reproduced in the accompanying woodcut.

My operation proved in every way unsatisfactory. It was more difficult to perform than had been anticipated, and the result was an utter failure. It was performed as follows: An incision was made along the centre of the *dorsal* surface of the *middle* finger from one end to the other, and a flap, including integument and subcutaneous tissue, dissected up laterally as far as the middle of the *ring* finger. Another and similar incision was made on the *palmar* surface of the ring finger, and the flap dissected half-way across the *middle* finger. The flaps were then held back, and the remaining tissue forming the bond of union between the two fingers divided. The union at the ungual extremity was osseous. The flaps were then adjusted; that dissected from the dorsum of the middle finger being wrapped round the palm or surface of the ring finger, and *vice versâ*.

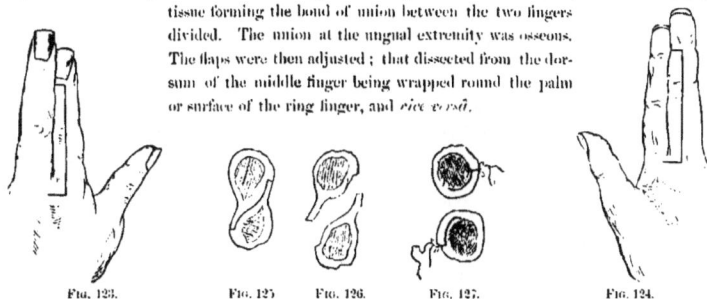

Fig. 123.　　Fig. 125.　Fig. 126.　　Fig. 127.　　　　　Fig. 124.

They did not adapt themselves as had been expected. The flap raised from the dorsum could be made to touch the line of integument on the palmar surface of the other finger by some stretching, but the palmar flap being tough, thick and inextensible, could not be made to meet by nearly a quarter of an inch.

The edges, however, were brought together as well as possible and a simple dressing of lint applied. A piece of rubber tubing was then passed through the septum between the middle and index finger, about half an inch above the normal commissure, and attached to a band round the wrist, to establish a permanent opening, and thus form a basis for future proceedings on the thick web remaining from the unsuccessful operation in infancy.

June 9th. Since the operation the fingers have been going on from bad to worse. No healing took place, but everything sloughed, until at one time it looked as though the boy would inevitably lose his hand.

To-day the second and third phalanges of the middle and ring fingers were removed, as they only hung by shreds of necrotic tissue. The attempt to make useful fingers out of useless ones had of course failed, and the only thing left was to prevent further sloughing and save the hand.

June 20th. The healing process has been going on slowly since the last report, and the wounds now present a healthy appearance, and but a small surface remains to heal over. Patient meantime has been kept on iron and nutritious food.

[121]

The hand is more deformed and useless than before the operation. The first phalanges, which remain, have reunited, and there is now a short, awkward-looking stump between the index and little fingers, which perhaps would be better removed, but the patient says he has had enough of surgery for the present. He was discharged July 4th.

The miserable result of this case was a bitter disappointment to me. The operation, studied in description and diagram, seemed so promising, and the results of mere division in these cases, as illustrated in the history of this very one, are ordinarily so unsatisfactory, that I had no hesitation or misgivings in undertaking a proceeding that seemed easy and certain to succeed.

But the inextensible nature of the integument, its close connection to the parts beneath, and the impossibility, from the very nature of the case, of making the flaps larger than the surface they are intended to cover, and thus allow for shrinkage, proved to be practical and fatal obstacles to these flattering anticipations.

From the experience gained in this case, I do not think that the operation is one to be recommended, and I should never undertake it again.

Indeed, I will go further, and say that it should be utterly discarded from surgery, and buried in the limbo of obsolete procedures. Division of the junction merely with rigid attention to the after-dressing, even if it sometimes yields very imperfect results, is far better than running such risks as the entire destruction of the fingers, or even the whole hand. The establishment of a permanent orifice at or just beyond the situation of the normal commissure, previous to division of the bond of union, would no doubt prove a useful procedure. Of course, that commoner class of cases, where there is merely a thin web of skin, are comparatively easy to deal with. But as I intend to

treat this whole subject *in extenso* in a future publication, I shall say no more about it for the present.

Of the other two cases of which I present illustrations, I have no history whatever to offer. The parties came under my notice quite accidentally. In each case it is the right hand which is malformed, having only two fingers and a thumb. The missing fingers are in each case the index and minimum. In one case the fingers present, middle and ring, are closely united throughout, and the curvature produced by this blending of two fingers of unequal length is well shown; the patient declined surgical interference, although the usefulness of the hand was much impaired. In the other case the fingers are perfectly distinct and well formed, and the hand quite useful. Both of them were laboring men; the other hands were natural; neither knew of any malformation among their relatives.

As far as I know, such cases as these are very rare.

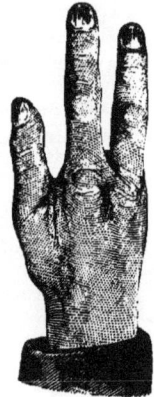

Fig. 128.
Fig. 129.

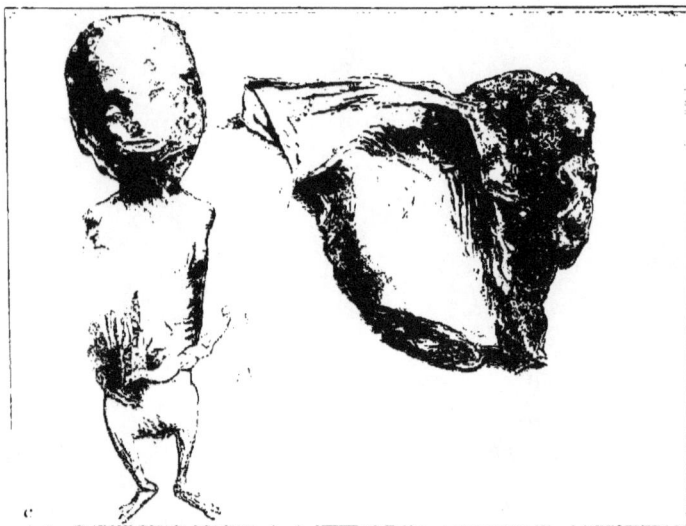

XX. a, HYDROCEPHALUS PEROPUS.   b, ANENCEPHALUS, HERNIA UMBILICALIS.
c, ABRACHIUS PEROPUS.
(Cases of Prof. G. J. ENGELMANN.)

A TERATOLOGICAL CONTRIBUTION.

BY GEO. J. ENGELMANN, M.D. OF ST. LOUIS,

Professor of Obstetrics in the Post Graduate School of the Missouri Medical College, St. Louis.

Although Teratology has advanced with the progress of embryology, it still lags far behind its sister science and may be said to be as yet in its infancy, the tardy development of this science being due to a superstitious dread and a feeling of abhorrence for monstrosities and deformities of all kinds, innate to man, common to all people and all civilizations, more fully developed the more primitive the people, yielding to a certain extent to civilization, but completely overcome by advanced science only.

The instinctive dread of monstrosities is so great among primitive people that the new-born child, if in any way seriously deformed, is at once put to death and done away with. This superstition among some peoples even extends to twins, and not only does the new-born infant suffer, but even the innocent mother is exiled or punished in other ways. Among the more tolerant tribes the more slightly deformed are indeed permitted to live, but they are endured merely as objects of pity; they nourish and clothe themselves as best they can, keeping out of sight of their people. So universal are these customs, more or less stringent, that so close an observer as Alexander von Humboldt makes the statement that deformities are extremely rare, if ever occurring, among primitive peoples. The fact is that few are permitted to live, and those who do live exist as best they can in solitude and exile. The Romans, while avoiding them, venerated them, as being gifted with peculiar powers. Cicero speaks of them: De Divinatione, I.: "Monstra, ostenta, portenta, prodigias nominantur, quia monstrant, ostendunt, portendunt et praedicunt."

During the middle ages they were again looked upon as products of hell and Satan, as it was believed that God could not form such horrible creatures, and this, of course, prevented the study of these unfortunate beings. They were regarded as a *Signum malioninis.* The remarkable oddity of form and peculiarity of shape often exhibited led to the most remarkable deductions, which were greatly furthered by fabulous descriptions and truly monstrous illustrations; and to these wanderings of imaginative minds was the science of teratology confined up to the end of the seventeenth century. The authors approved of the barbarous laws of the Greeks and Romans, condemning monstrosities to death, and it is with great misgivings that some of the more advanced admitted that giants, dwarfs and children with supernumerary fingers and toes should be permitted to live, but demand that even these must be raised far from human intercourse.—Riolanus (De Monstro nato Lutetiae, 1605).

Not until the beginning of the eighteenth century does scientific research begin, and since then, especially with the science of embryology, has the study of teratology gradually developed. No satisfactory conclusions have as yet been reached. Almost every author follows his own classification and his own views, and much work must be done before these can be made to harmonize.

Foerster truly states that morbid processes in the ovum result very differently, according to the degree of development during which they occur, and consequently separates the resulting monstrosities into two grand divisions.

First. Morbid changes similar to those which result from the same diseases after birth are such as appear after the beginning of the fourth month, when all parts are formed and merely increase in size.

Secondly. The totally different changes which occur before this period, when one or more parts are not yet formed or are still developing; these are the real monstrosities, and if we think but a moment this must teach us how ridiculous are the supposed effects of maternal impressions, often occurring when the affected parts have long been perfectly formed. He establishes three classes:

First. Supernumerary or oversized members.

Second. Defective.

Third. Abnormal shape or changes in the germ or embryonic development.

I merely state these to show how unsatisfactory they are, as one single monstrosity may appear under all of these classes, and repeat that I will attempt neither classification nor explanation, but will merely describe these three monstrosities with the hope that they may prove useful to others. The gathering of material and description of individual cases will serve the specialist in his work, and it is with a view to this that I add my mite by the brief description of the following somewhat unusual cases which lately came into my hands:

a.—Hydrocephalus Perquus.

This female child, the result of a premature delivery in an unfortunate girl, a primipara, is thirteen and five-eighths inches in length; the head measures three and seven-eighths inches; the body five and seven-eighths inches; the lower extremities three and seven-eighths inches. The peculiarity consists in the combination of hydrocephalus with deformity of the extremities, in addition to which we have the supernumerary toes and fingers.

The hands have five fingers each with a morbidly changed sixth, which appears more like a tumor attached to the little finger; on the right foot we find six well-shaped toes, so that it is difficult to say which is the supernumerary; whilst upon the left foot the little toe is so deformed as to make it evident that the addition is in this member. The deformity of the lower extremities in connection with the hydrocephalus and other morbid changes would undoubtedly lead us to presume that they had been caused by faulty innervation and not by local influences, such as pressure due to the position of the child in the womb, as is often supposed, and possibly the actual cause in many instances. In this case, although the body of the child very easily and perfectly admitted of being replaced in the position it had *in utero*, the very striking deformity in legs and feet could hardly be thus accounted for, but must be referred to pathological changes in the nerve centres, which unquestionably exist; I believe that this specimen is one of unusual importance, as it seems conclusively to prove the dependence of certain deformities, especially of legs and feet which are so frequently ascribed to position and consequent compression, to a pathological condition of the nerve centres.

I would by no means deny that deformities, above all pes valgus or varus, may result from compression or other local influences, more particularly in a feeble organization, but the advocates of each theory are too sweeping and general in their deductions, ascribing all deformities to one cause, whereas to my mind we here have a striking example of the effect of faulty innervation. The validity of one cause does not exclude that of the others.

M. C. Darest (in a recent paper, *Académie des Sciences. Compts rendus de la Séance du 23 Jan. 1882*) gives some striking examples of the development of monstrosities by posture and by compression of the embryo in consequence of tardy development of the amnion, which acts as the compressing agent. This fact he has observed in birds' eggs, but also in the embryo of mammals; and he believes that the deformities of the feet in the human fœtus are mainly due to this cause.

*b.—*Anencephalus, Hernia Umbilicalis.

A female fœtus eleven and three-eighths inches in length; the face two inches; body four and ten-sixteenths inches; lower extremities five and seven-sixteenths, whilst the feet measure two and one-quarter inches; the head, as is usual in these cases, sits directly upon the shoulders, with no neck. The cleft in the spine extends to what would probably be the first dorsal vertebra. Some little fluid, no cerebellum whatsoever in the small sac pendant from the spinal aperture; the anterior portion of the skull covered by a thick scalp with some hair. The peculiarity lies mainly in the abdominal organs, these are all, with the exception of the kidneys and pelvic viscera, contained in the umbilical sac, which has a circumference of seven inches, extending three inches from the abdomen. The navel-string is inserted in the lower third of the sac, which has somewhat the appearance of the integument and passes in this to the abdominal wall below. The circumference of the sac at its base is a little over two inches; it is a delicate semi-transparent membrane, very much like the chorion; in it lie the entire liver, spleen, stomach and intestines; the ligamentum suspensorium hepatis is attached to the sac. The abdominal cavity is very small and completely filled by an immense horse-shoe kidney, the left lobe of which is over an inch in length and twice the size of the right; both, however, extending up to the diaphragm. The descending colon passes down in the centre of the body over the bridge of the horse-shoe kidney and the promontory, in the hollow of the sacrum behind the uterus. The pelvic viscera are all present, but, like the sexual organs, somewhat backward in their development; the Fallopian tubes with their fimbriated extremities are very marked, as are also the ligamenta ovarii, while the ovaries themselves are simply recognizable as a thickening of the tissue. The external sexual organs are fully developed, with a double vagina.

[125]

A TERATOLOGICAL CONTRIBUTION.

c.—Abrachius Peropus.

Female fœtus, five inches in length; head one and seven-eighths inches; body two and three-eighths; lower extremities three-fourths. The head is large, face small; external ear rudimentary, situated lower than ordinary from the tip of the nose to the line of the upper lip; the breadth of shoulders is one and one-half inches. The arms are entirely wanting; the apparent stump is owing to a development of the condyloid process of the scapula; instead of the glenoid cavity we find a round, smooth, cartilaginous protuberance appearing like the caput humeri. The scapula and clavicle are otherwise well formed. The pelvis is small. The leg is formed by a single bone which has a large head, and is attached to the pelvis loosely by numerous ligaments. The head of the bone has more the appearance of that of the tibia than that of the femur. The bone itself is more like a femur than a tibia. It is round in shape and is directly attached to the long, narrow foot which has three metatarsal bones and three toes, corresponding to the large toe and the two adjacent ones. The fibula and two toes, which may be said more or less belonging to this, is wanting. The afterbirth also offers various peculiarities. The cord is very much twisted; the fœtal extremity attached low down, less than half an inch above the pubes and one and seven-eighths inches below the neck. The placental extremity enters the membranes fully half an inch from the border of the placenta, there being really no placental attachment whatsoever. The placenta is three and one-half inches in length by two and one-half inches in breadth, and is peculiar in the irregularity of the formation of its lobes. The condition of the funis is deserving of some attention; whether influencing the irregular development of the fœtus in any way or not I dare not say, but certain it is that a thinned or twisted cord necessitates an insufficient supply of blood, as the circulation is obstructed, if the vessels are of sufficient size [in many cases they are contracted], so that even without additional mechanical obstruction they are unable to supply the demand. In most cases the development of the embryo is thereby retarded; it dies, and abortion follows, either of an incompletely developed ovum, or of a mole if the ovum is retained in utero a certain length of time after death of the embryo.

These conditions I have frequently observed, and I am inclined to attribute the cause of the miscarriage in this case to the death of the embryo from insufficient nutrition through the obstructed vessels of the funis, yet I doubt whether this will in any way explain the deformity, although it is one of retarded or incomplete, not of active though perverted development.

Unfortunately, I can obtain no history of any of these monstrosities but the first. We certainly cannot account for this deformity by the fact of the mother's anxiety or the troubles of an unfortunate girl, and no history would assist us in explaining the peculiar formation of the various organs in the other two.

APPARATUS FOR TREATING FRACTURE OF THE PATELLA.

BY J. S. WIGHT, M.D.,

Professor of Operative and Clinical Surgery at the Long Island College Hospital, Brooklyn, N. Y.

In treating a fracture of the Patella it is desirable to bring and keep the fragments in as close apposition as possible. There are two principal obstacles to attaining this result: (1) A firm blood clot on the broken surface of the fragments—as I have demonstrated by post-mortem examinations; (2) The contraction of the muscles of the quadriceps extensor.

Some years ago I devised an apparatus for treating fracture of the patella, and in 1880 I reported the treatment of three cases by this apparatus. (See No. 1230 *Medical and Surgical Reporter.*) My colleagues and myself have since treated several cases of fracture of the patella by the same apparatus.

The essential parts of this apparatus are represented in the accompanying drawing, Fig. 133. First, a single inclined plane is made by firmly joining two pieces of board at a right angle: one is a bed-piece, and the other is an upright-piece, as seen in the diagram. The bed-piece is bevelled where it goes under the thigh. Two quadrangular bars are hinged to the upper end of the bed-piece. These bars are long enough to move up and down on the sides of the upright-piece, and are sustained by a bandage going over a hook on the outside of the upright-piece. A piece of muslin or canvas, is fastened to the movable bars, making a kind of hammock on which to put the injured limb. This can be seen in the drawing.

Second. The upper ends of two quadrangular pieces of adhesive plaster are cut into strips and applied obliquely from below upward—in one case inward, in the other case outward—on the dorsum of the thigh, over the quadriceps extensor. The lower ends of these extending straps are properly attached to cords that play over wheels fastened to the sides of the upright piece, and that have weights suspended from their distal ends. A roller bandage, beginning on the upper fragment of the patella, is passed around the thigh upward for a sufficient distance to prevent the extending adhesive straps from coming off on account of the tension of the weights, which may vary from three to six pounds each.

Third. The lower fragment of the patella is attached to the tibia by an inextensible tendon, and can be kept in place by means of a small sand-bag resting on it, as in the diagram. When the lower limb is in position, on the single inclined plane, the sand-bag will act on account of the force of gravity. It may be necessary to fasten this sand-bag on by adhesive plaster, in order to prevent it falling off sideways.

This apparatus has the following advantages :

1. The force of gravitation is utilized for keeping the fragments in apposition, and is applied in such a way as not to cease acting.

2. The patient can move about in bed to a considerable extent, and not interfere with the proper action of the retentive forces ; and this affords great relief, as I have more than once seen.

3. The limb will not be excoriated by its contact with the hammock on which it rests.

4. There will be no material interference with the blood supply to the fragments of the patella, and so delay union.

5. The leg will be left free, so that passive motion can be made at the ankle-joint from time to time : also passive motion can be made at the knee-joint in such a manner as not to interfere with union, and so as not to make anchylosis so imminent.

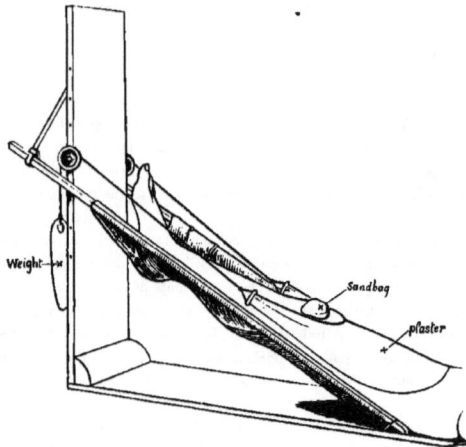

Fig. 133.

XXI. ENCHONDROMA OF THE HUMERUS.
(Case of Dr. H. H. Dawson.)

ENCHONDROMA.

BY W. W. DAWSON, M.D.,

Professor of Surgery, Medical College of Ohio, Surgeon to Hospital of the Good Samaritan, Cincinnati. Ohio.

J. D. H., aged 32, the subject of this tumor, has a fair family history: his parents died in mid-life; father, of pneumonia; mother, of some obscure affection called "liver trouble." He has one sister living, at 25, in robust health; lost four during infancy and childhood. His general health has been good, the tumor giving him more inconvenience than pain.

The early history of this tumor is indefinite; he learned from his mother that on one occasion, when he was about nine months old, he cried on being lifted by the left arm, and that from that time on, a gradual development was noticed; previously, he had been a healthy, plump baby. He says, "I first remember it when it was about the size of a baby's head, it was somewhat egg-shaped, and painless; I played with other children, and was only inconvenienced by its size. Once, when a little boy, I hurt it, and was laid up for several months."

He has been selling newspapers at the wharf for twenty-one years, ever since he was eleven years old, interrupted only by a couple of trips to New Orleans, on a steamer, as "deck-sweeper." He bears exposure well, especially over the tumor he wears light covering. Weight is 135 lbs., height 5 ft. 7 in., measures around the chest 31 inches, waist 28¾ inches. On the affected side the hand is muscular, and he has considerable strength in the arm; with it he can pull, but he cannot lift. The dynamometer shows, right hand 85, left 48. The left arm is about one-half inch shorter than the right, and nearly an inch less in circumference.

Size of Tumor.—Circumference at most prominent part 38¾ inches; greatest length—from above downwards, 30¼ inches.

Shape.—As will be seen by the illustrations, the tumor is conglobate, the lobes being irregular in size and outline. Viewed from the rear, four great divisions appear: these are again subdivided into smaller ones. Upon the anterior aspect the lines are not so deeply cut. There is a kind of ridge on the summit, and at the extreme point of this is situated the most symmetrical lobe: this is the seat of traumatic ulceration.

Consistence.—The larger divisions are dense, though elastic. They impart the impression of thick walled, tensely filled cysts. The smaller nodes are more compressible, and give an impression of fluid beneath a dense covering.

[129]

ENCHONDROMA.

Traumatic Hemorrhages.—The first severe injury which the tumor received did not result in hemorrhage; at each subsequent trauma, however, there has been more or less loss of blood.

Spontaneous Hemorrhages.—During the last three or four years he has had frequent spontaneous hemorrhages. They occur mostly at night when he is in bed; it is usually more annoying than an active flow. He applies cloths with slight compression, never calls on a physician for assistance, although the wasting sometimes keeps up for several days. During the past few months—the fall and winter—the loss has been slight, but well-nigh constant. The waste is telling on his health. Before the holidays he resorted to the Good Samaritan Hospital for a few days' rest; absence from his business has been a very unusual occurrence.

Ulceration.—The damaged point has remained in a state of chronic inflammation ever since the first hemorrhage. The breach presents no appearance of malignancy; it has the aspect of an indolent ulcer. The low vitality of the part renders complete repair next to impossible.

Growth.—This tumor has had a continuous growth from the initiative. The patient thinks that from year to year he can notice a gradual increase in size.

Painless.—Except when it has been injured, it is absolutely devoid of pain.

Blood Supply.—The large venous trunks ridging the surface of the nodules, although compensatory, indicate a liberal blood supply, and taken in connection with the hypertrophied condition of the parts around the shoulder rendering the medium of attachment with the body very large, preclude all possibility of surgical relief, the circumference of the arm at the axilla being 25½ inches.

Chondroma, Osteo-Chondroma, Enchondroma. Osteo-sarcoma, Osteo-chondrophyte, are but a part of the terms which have been used to designate cartilaginous growths, hyperplasie which at one period in their development are composed in whole or in part of cartilage; they are all, to some extent, indefinite.

Osteo-sarcoma is appropriate so far as it applies to growths which are partly osseous and partly carneous, but beyond this it is as unsatisfactory as that ancient, that vastly-including cognomen, sarcoma. Eechondroma by some is applied to those cartilaginous products which spring from the surface of bone, while enchondroma is confined to those which originate in the interior.

This large mass is composed of a number of lobules differing in size and consistence, and covered by a dense layer of connective tissue capsule. The surface, normal in color, is made irregular by the passage of large blood-vessels, superficial veins—improvised for maintaining a compensatory circulation.

In looking for the genesis of this tumor, we are led to the periosteum; it is that part of the humerus covered by this membrane which has disappeared, the articular extremities remain intact; hence pronation and supination are free, and flexion can be made until arrested by the overhanging mass. The cartilaginous caps upon the ends of the humerus limit the attachment of the periosteum, and limit also the destruction of bone and the formation of this chondroid tissue,—between these points the degeneration is confined.

Our knowledge upon this kind of pathological substitution is very meagre indeed, a process by which one tissue is removed, and another formed to occupy its position.

[130]

A CASE OF

INTELLECTUAL MONOMANIA, WITH MENTAL DEPRESSION.

BY WILLIAM A. HAMMOND, M.D.,

Surgeon-General U. S. Army (retired list), Professor of Diseases of the Mind and Nervous System in the New York Post Graduate Medical School, etc.

This phase of mental disorder is not to be confounded with the emotional form of insanity known as lypemania or melancholia, with which, though entirely distinct, it has naturally many relations. It is the *monomanie triste* of Marcé, and as this author has pointed out, is characterized by the fact that although the patient has fixed delusions of a melancholic character which influence him in his actions, he can nevertheless reason well in regard to other subjects, and is often able to conduct himself with entire propriety in all the relations of life outside of his own particular erroneous beliefs. In melancholia, on the other hand, the emotions are involved to an extreme degree; the false conceptions which exist often assume entire control of the mind and render the individual altogether incapable of the systematic performance of rational acts, whether they are or are not connected with his delusions.

Without at present going into a detailed description of intellectual monomania with mental depression, I desire to say a few words relative to what is sometimes one of its most striking manifestations, and that is the condition known as the *delirium of persecution*. This has attracted the serious attention of alienists not only on account of its pathological relations, but of its importance to medico-legal science.

Generally, this state begins with illusions and hallucinations, which for a time may be stren-

[131]

uously resisted by the individual, but which usually eventually obtain a complete mastery over his reason. The sense of hearing is that which is generally the seat of those false perceptions, which appear either as vague uncertain sounds or isolated words, or as well-defined and entirely coherent sentences. These are in the form of threats, or warnings, or advice as to the best way of escaping from imaginary enemies or dangers. The sense of sight is not so frequently affected, though occasionally the patient sees a policeman or other person in search of him in every one who looks at him a little closely. In order to escape from these imaginary enemies he makes complaint to the officials or seeks safety in flight, or may even proceed to the extent of perpetrating suicide or homicide. Sometimes the individual labors under the delusion that organized bodies of men have banded together for the purpose of destroying him or of inflicting severe bodily injury upon him. These may in his imagination be the whole police force, or the clergy, or the medical profession, or the masonic fraternity, or the members of some one nationality. A patient of mine was sure that all the clergymen had entered into a conspiracy to "pray him into hell." He went to the churches of all religions denominations to hear what the preachers had to say about him, and discovered adroit allusions to himself and covert invocations to God for his eternal damnation in the most harmless and platitudinous expressions. He wrote letters to various pastors of churches denouncing them for their uncharitable conduct towards him, and threatening them with bodily damage if they persisted in their efforts to secure the destruction of his soul.

Another was constantly dodging around the corners of the streets and hiding himself in doorways to avoid detectives, for whom he mistook all who happened to look at him with more than a passing glance, and who he conceived were seeking to arrest him on the charge of attempting to take the life of the mayor. "I never even saw the mayor," he would exclaim with tears in his eyes, "and God knows I never wished him any harm, and yet these scoundrels are endeavoring to imprison me for shooting a pistol at him. There's another of them!" and instantly he darted down an area to hide till a bland-looking old gentleman whom he took for a disguised detective had passed. "That man," he continued as he emerged from his place of seclusion, "is the sharpest one of the whole lot. He looks seventy years old, but he's only twenty-five. His hair is a wig and his beard is false. I can go nowhere without just managing to escape. Of course he will catch me at last, and then I shall go to prison for life."

C. B.,* after separating from her husband and remaining absent six years, came to the United States from Ireland and there married again. Shortly afterward a daughter by her first husband came over, and then the mother seemed to realize for the first time that she had two living husbands. This idea seemed to be the exciting cause of her insanity, which first showed itself in unfounded suspicions that her daughter was leading an improper life. Hallucinations of hearing next supervened, to the effect that people were talking about her night and day. She imagined she heard a young man say that she was a bad woman, had stolen lands, committed forgeries, and was the mistress of a Mr. Welsh. Also heard him say that a play founded on her life was being performed at a theatre. She declared that people look crossly at her and point their fingers towards her. Was very positive about all she heard and saw, and said her opinion could not be changed if all the circumstances should be denied by

* From Dr. Parsons' MS. Notes of Cases in Blackwell's Island Asylum.

the persons whom she thought had spoken about her and pointed at her. This patient remained in the asylum for several years in about the same condition as when she entered it.

Delusions of poisoning are very common with these people. A man from Brooklyn only a few days ago came to visit me, and having to wait his turn in the reception room, sent in a note to the effect that he had been poisoned by a man with whom he had dined a short time since, and that he would not wait as the poison was " working on him." I had treated this patient a year previously for similar delusions and he had entirely recovered and had resumed his business, that of a shopkeeper. Some time before I first saw him he had been an inmate of the Insane Asylum at Flatbush. I sent for him to come into my consulting room, and, to quiet him till I could attend to him, poured out a dose of the fluid extract of coca and requested him to swallow it. He took the tumbler into his hand and looking at it for a moment, set it down hastily and rushed from the house, exclaiming, " You are as bad as the rest of them ; just as bad as the rest !" A few days afterward he visited my son, Dr. Greene M. Hammond, with a similar story of poisoning, but left hurriedly while preparations were being made to examine him.

It is not at all uncommon for the victims of delusions of persecution to imagine that they are being acted upon by some occult influence or by some one or more of the forces of nature, as heat, magnetism, or electricity. " Spells " are laid on them by certain individuals whom they know, or by invisible persons who only make themselves known by their speech. In one case that was under my charge the patient, a stationer doing business in this city, had the delusion that unknown enemies—freemasons—were acting on him by electricity, which they sent into his brain through the top of his head, by powerful batteries which they had in their lodge-rooms.

In another, a woman who kept a small shop in the Bowery, and who came to my clinique at the Bellevue Hospital Medical College for the purpose of getting relief, declared that all the iron railings and railway tracks had been charged with electricity in order to injure her, and that whenever she touched one of them, or even came near them, she received a severe shock. A case of a like character is cited by Semelaigne.[*]

Very slight causes are sometimes sufficient in a patient suffering from intellectual monomania with depression to excite hallucinations which have been for some time absent. Poterin du Montel [†] cites the case of a woman who had become melancholic, lost sleep, and had pains in the head and bleeding from the nose in consequence of some insignificant family disagreement. She contracted the delusion that her sisters, who were in reality devoted to her, had conspired to injure her. She also had illusions and hallucinations, saw a black head and heard voices speaking against her. The mere opening or shutting of a door, a step on the floor, or the slightest sound, was sufficient to excite these hallucinations.

A somewhat similar case was at one time under my observation in which the patient, a lady thirty years old, whose mother had died insane and who was herself of a strongly-marked nervous temperament, suddenly became affected with hallucinations of hearing, by which she was told that her

---

[*] " Du Diagnostic et du Traitement de la Melancholie." *Mémoires de l'Academie Impériale de Médecine.* Tome XXV., p. 235.

[†] " Études sur la Melancholie," etc. *Mémoires de l'Academie Impériale de Médecine.* Tome XXI., p. 462.

INTELLECTUAL MONOMANIA, WITH MENTAL DEPRESSION.

servants had entered into a conspiracy to burn the house and her with it. Although she never had any hallucination of seeing the persons from whom the voices were supposed to come, she was quite sure that they came from real individuals concealed in various parts of the house or under the steps of the houses she passed in the street. Night and day while awake she heard the voices. Finally, the continuity of the hallucinations ceased, but the delusion remained, and she was constantly watching her servants, frequently changing them, and invoking the aid of the police in order to insure her safety. But if at any time she heard a very loud noise, such as the rumbling of a heavy wagon in the street or the explosion of a blast, the hallucinations at once returned.

The foregoing description is perhaps sufficient to point out some of the chief features of intellectual monomania with mental depression so far as they are connected with the delusions of persecution which are often present. In the next place I desire to present to the society the history of a case of universal interest, in which such delusions existed.

On October 10th of the present year, a man dressed in ragged and paint-soiled clothes walked with a quick, nervous, and uncertain gait along the south side of Fourteenth Street, between University Place and Fifth Avenue. His unkempt and disorderly appearance attracted attention, and as he hurried on, his head bent upon his chest and his eyes staring wildly from under a slouched hat at the passers by, many avoided him, instinctively forming the idea that he was contemplating mischief.

The street was crowded, mostly with women, but the man pushed the people aside, at the same time muttering some unintelligible words, and assuming an angry expression. Suddenly he drew from the breast pocket of his coat a pair of sharp-pointed compasses such as are used by carpenters, and began to strike right and left at the people around him. In a few seconds he had struck and wounded three women who were near him. The rest scattered in all directions, the man pursuing some and striking at them with his murderous weapon. Three other women were wounded, and then the man was seized by several men, and a policeman coming up, he was disarmed and taken into custody. It was then ascertained that he was a Frenchman, named Ernest Dubourque. All the women recovered except one, who died of secondary haemorrhage several days after the infliction of the wound.

When questioned relative to his reasons for his murderous conduct he either gave unintelligible answers or refused to reply, alleging that he did not understand English. He frequently rubbed his head with his hand and complained of pain. It was then recollected that for several years past the man and his father had been perambulating Broadway carrying signs on their backs stating that they had been defrauded out of a large fortune by the United States government, and demanding the restitution of the money of which they alleged they had been robbed. The only ground for this conduct were the facts that a brother of the old man had some twenty years ago died in California, and it was supposed he had left a great deal of gold behind him. The father and son never, however, took any pains to ascertain the truth or falsity of the report, but contented themselves with charging the government with having appropriated their legacy to its own uses. During all this time they were regarded as harmless lunatics. Less than a year ago the old man died, and then the son was compelled to walk alone. Occasionally he did a little work as a painter, and appears to have suffered at times from lead colic. He continued to talk of his grievances and told improbable stories of his acquaintance with great men in France, and his intimacy with the nobility. A few months since, while painting a

[134]

ship, he assaulted one of the workmen whom he accused of robbing him, and was, in consequence, sent to the prison on Blackwell's Island. On his release he declared that he had been incarcerated by the men who had stolen his inheritance. Subsequently he stabbed a policeman, was again arrested, but was, strange to say, discharged from custody and allowed to go at large on the ground of insanity. A few days after his attack on the people in Fourteenth Street I examined him at the Jefferson Market prison, in which he was confined. He kept his eyes fixed upon the floor, only raising them when I asked him some question, and then only for a moment. His expression was dull and stupid, his extremities cold, his pulse 65 and weak. His face was somewhat swollen, and his forehead was marked by the deep transverse lines so often seen in lunatics of his type. In answer to my inquiries he said that he was 37 years old, and that he had frequent attacks of pain in the head and vertigo. At first when I asked him why he had stabbed the women, he shrugged his shoulders and muttered something about their having been talking against him. He refused to tell what they had said, remarking that he did not understand English. When the question was put to him in French he shrugged his shoulders again, but said nothing. I told him that one of his victims would probably die (she died at just about that time, though I did not then know the fact), and that if she did it would probably go hard with him. To all of which he made no reply, but looked the picture of perfect indifference. He then wrote his name, " Ernest Dubourque," on a piece of paper which I gave him for the purpose.

On other things than his crimes he talked glibly enough, both in French and English, but as soon as they were mentioned he shrugged his shoulders and said he did not understand. Finally, he declared that he knew nothing about the matter, and did not know for what he was in prison.

My interview led me to the conclusion that Dubourque was a lunatic, and sufficiently shrewd to pretend ignorance of his conduct. At first his cunning was not so well marked, for he admitted that he had stabbed the women because they had been talking about him, but as I questioned him further his mind seemed to awaken, partially, to the idea that my inquiries were hostile to him, and that to answer them would get him into further trouble.

The clinical history of Dubourque, aside from his personal appearance and manner, is of such a character as to show that he is the subject of intellectual monomania with mental depression. His insanity, as well as that of his father, appears to have been developed by disappointment that they had not obtained a fortune by the death of the California relative. So strong was the idea—for which it appears there had never been the least foundation—that he was rich, that when he died they contracted the additional illusion that he had made a will in their favor, that the will had been destroyed by the government, and that his property had been turned into the United States treasury. For several years Ernest Dubourque and his father had walked Broadway with placards on their backs, wishing by this means to make the world acquainted with their supposed wrongs. It was a clear case of the *folie à deux*, to which attention has within a recent period been so pointedly directed. The death of the old man appears to have aggravated Ernest's condition. He became aggressive, he quarreled with his fellow-workmen under the idea that they were attempting to cheat him, and stabbed a policeman for some fancied indignity. Finally, he committed the numerous assaults resulting in one death for which he is now in prison. Under the new code it is difficult to see how he can escape punishment for his crime. Under that code, in order to avoid responsibility, it must be shown that he was laboring under such a defect of reason as either not to know the nature and quality of the act he was committing, or not to know that it was wrong. Morbid impulse will not therefore longer be an excuse, nor will a delusion unless such delusion would, if true, justify the act. The most that Dubourque could urge in extenuation was that the women were "talking" about him. He certainly knew the nature and consequence of his acts, and knew that they were contrary to law. Undoubtedly, therefore, he ought to suffer punishment. And I would ask, in conclusion, whether or not the corporation that allows a lunatic to go at large when he has once been acquitted of a murderous assault on the ground of insanity, cannot be mulcted in damages for its open and flagrant disregard of the public safety. The disposition shown by juries to pronounce lunatics sane is doubtless the natural outgrowth of the asylum abuses, which are such a blot upon our national reputation. It might, however, have been expected that sheriff's juries, selected as they are presumed to be from a superior class of the community and endowed with intelligence above that of the common jury, would have been superior to the influence of such a factor. But experience shows that they are not; for within a few days past they have pronounced a man sane and competent to manage his business who is in the initial stage of general paralysis of the insane, and who is quite certain, unless soon restrained by force, to ruin his affairs and perpetrate some act of violence. Here the claims of the man to protection against himself and the still more imperative claims of his family were with that restricted vision peculiar to narrow-minded persons, unheeded in presence of the more obvious fact that they were discharging an alleged insane person from an asylum. The alienists and neurologists who, after thorough examination, testified that the man was the subject of general paralysis of the insane and that he was hopelessly incurable, can well afford to let time act as the arbiter between them and ignorance. In the meantime, the person, the family, and the property of the insane individual suffer till such time as his lunacy becomes so palpable that even sciolistic commissioners, so-called experts, and sheriff's juries can no longer doubt its existence. Even as these lines are being written the person in question, with maniacal triumph, has published an advertisement in the daily newspapers which is of itself sufficient evidence of his lunacy. Besides, he has made a murderous menace with a loaded pistol against a man who was doing some work about his house, and who did not do it quickly enough to suit him. Surely it would seem that something more than the verdict of a sheriff's jury is necessary to protect the community from such people.

TRICHOPHYTOSIS BARBAE.

(*Synonyms*—Tinea Barbae, Sycosis, Mentagra, Barber's Itch.)

BY HENRY G. PIFFARD, M.D.,

Surgeon to Charity Hospital, Blackwell's Island, etc., New York.

In former times, and up to the early part of the present century, two, possibly three distinct diseases were included under the names Sycosis or Mentagra. Gruby first threw light on the true nature of one of these by the discovery of its real cause. Later, Bazin studied it more fully, and to him and succeeding writers we owe our present knowledge of its nature and treatment. The affection may be illustrated by the following cases:

J. D., forty years old; had always been in the habit of shaving himself. On one occasion, however, he was obliged to avail himself of the services of a barber. A few days later, he noticed two red pimples in the region of the moustache. These were followed by others on the cheeks, and were accompanied by slight itching. The papules increased in size, and appeared to heal in the centre, until in a short time he found two pretty well defined rings on the left cheek, and three on the right. The rings increased in circumference until they united, forming as it were one large irregular ring on each side of the face. The disease had existed about two months when he came under observation. At this time he presented the appearances shown in the photograph. The eruption consisted of a reddened, raised, irregular ring on the right cheek, with distinct margin.* On the left side the circle was less uniform, showing more clearly the fact that it was made up of two primary rings. Beside these there were a few scattered papules, and two or three larger tubercles. He had been subjected to various treatment. I directed that the hairs should be carefully and thoroughly removed with the epilating forceps, and that a solution of corrosive sublimate, two grains to the ounce, should be applied morning and night. Internally he was directed to take a quarter of a grain of Calx Sulphurata thrice daily.

S. C., thirty-three years old, was shaved by a hotel barber in Boston. A few days afterward he noticed the commencement of the trouble. An eruption appeared, accompanied with considerable inflammation and pain. It increased rapidly in extent, quite a number of rings having formed and united. The inflammation was severe, and accompanied with considerable sero-purulent discharge, which matted the unshaven hairs together. The physician under whose care he came, fully appreciated the nature and requirements of the case, and advised epilation and the use of parasiticides. On

* In the Artotype illustration the photograph is reversed, the eruption appearing upon the *left* cheek.

attempting epilation the patient rebelled, declaring that the pain was greater than he could bear; and, in fact, actually refused to submit to it. Under these circumstances, his physician referred him to me. On examination I found a high grade of inflammation, with sero-purulent exudation, involving nearly the entire region of the beard. The parts were extremely sensitive, the slightest traction on a hair causing the patient to cry out. Finding that epilation was for the present, at least, impracticable, I ordered Calx Sulph. in doses of one-tenth grain twice daily, together with the following ointment:

℞
　　Fl. Ext. Stramonii recentis . . ℥ i,
　　Ungt. Hydrargyri Ammoniati　℥ i,
　　M

to be applied morning and night. Under this treatment, maintained with little variation, he rapidly improved, and was discharged from further treatment in about seven weeks.

L. A., twenty-five years old, and recently married, consulted me in January of the present year for an eruption that had appeared a week or ten days before, on his scrotum and thigh. On examination, I found four trichophytic rings on the thigh, one on the scrotum, and one on the penis. He stated that his wife was free from any similar trouble, and denied that he had had any other opportunity of acquiring it. I was puzzled to account for the appearance of the eruption in the regions mentioned, until I noticed a small spot on his neck. This, on inspection, proved to be a small scaly ring, in the region of the *pomum Adami*, which he stated had appeared shortly after the attentions of a strange barber, while on his wedding trip. I ordered daily frictions with the *Oleatum Hydrargyri*. A few days later he brought his wife, who exhibited a well-marked ring-worm, involving a portion of the upper lip, and the left ala of the nose.

The three cases related exhibit three phases of *Trichophytosis Barbae*, differing from each other however, in many respects, so far as appearance is concerned. In the first case we had well-defined lesions of a plastic character; in the second, a diffuse inflammation with abundant sero-purulent exudation; and in the third, the slightest possible degree of inflammation so far as the lesion on the neck was concerned. It will be noticed that the treatment differed somewhat in all three cases, in accordance with the difference in the lesions. The rationale of this will be better understood when we consider the natural history and etiology of the disease. We will first, however, say a few words about diagnosis.

DIAGNOSIS.—The question of diagnosis lies between trichophytosis proper, and one, or possibly two other affections. These are Eczema of the beard, and Sycosis idiopathica. Whether this latter affection really exists as a separate disease, is open to some doubt; the evidence in its favor, however, has been ably presented by Dr. A. R. Robinson of New York.

Be this as it may, there certainly exists a form of Eczema that occupies the same situations as *Trichophytosis barbae*, and from which it should be carefully differentiated. This can in most cases be readily accomplished from the history alone. In Trichophytosis we have the appearance of little red

[138]

points gradually enlarging and forming rings which later on, as the disease advances, run into each other with obliteration of the lines of junction. The inflammation accompanying this may be sufficient to induce the formation of tubercles and nodules of some size, which may remain as such for a long time, or undergo early suppuration. In *Eczema Barbae* we have a diffuse inflammation of a more superficial character with a dipping down or continuation of it into the hair follicles; the root-sheaths become swollen and loosened from the abundant exudation that forms in the follicle. The hairs, as a rule, are readily extracted, and come out with their root-sheaths adhering to them. The hair appears to have a thick white envelope surrounding its deeper extremity, in marked contrast to the hairs in trichophytosis, which are usually extracted free from the root-sheath, and often broken, leaving a portion of the root still within the follicle. The diagnosis may further be confirmed by microscopical examination of the hairs, which in the latter affection exhibit the parasite. Failure to find the parasite, however, must not be taken as absolute proof that the disease is not parasitic, as it does not necessarily involve all the hairs of the affected part. In the advanced stages, too, it may be absent, having already been destroyed by the efforts of nature, as sometimes happens. In the Sycosis of Robinson the inflammation is deeper, and commencing between the hair follicles, subsequently involves them.

ETIOLOGY.—As is now well known, the cause of Trichophytosis is a fungus termed the *Trichophyton tonsurans*, which having lodged on the skin, immediately makes the hair-root and follicle its principal abode. This fungus consists of exceedingly minute spores, together with a more or less abundant mycelium. The spores are in all probability the agents of contagion. After they have gained a foothold in the follicle and have commenced to invade the hair, commencing in the softer portions of the root, we find that they gradually work their way toward the surface, thoroughly infiltrating the shaft of the hair in their progress. The portion of the hair within the follicle, supported as it is by the follicle walls, apparently maintains its integrity, but as soon as the fungus reaches the portions of the hair-shaft that are without the follicle and unsupported, their destructive influence becomes more manifest. This is shown by rupture and fracture of the hair, the shaft breaking off at a very short distance from the surface of the skin. This condition is sometimes noticed in connection with Trichophytosis *barbae*, but much more frequently and distinctly in Trichophytosis *capitis*. This latter affection, however, is confined almost exclusively to children, never in our own experience having been encountered in adults. As noticed in our third case, Trichophytosis barbae may co-exist with other forms of the affection. The modes by which the disease may be contracted or conveyed are quite numerous. In the first place it is most frequently conveyed through the unclean implements of the barber. The razor has most generally been accused as the medium of communication; but, we think it more probable that the lather-brush is the offending instrument. The barber-shop, however, is not the only source of contagion; theoretically, it may be contracted from any form of ringworm. Personally, we have known of its having been contracted by a gentleman from his child, who was suffering from a ringworm of the scalp; in another instance, by a gentleman from a case of trichophytosis *genito-femoris;* and third, by a hostler from a horse, that he was in the habit of grooming. The occurrence of trichophytosis in the horse and other lower animals, has been specially noticed and studied in France.

COMPLICATIONS.—In persons predisposed to plastic or purulent inflammation, these features may become quite pronounced, while those who are subject to Eczema may develop it in connection with

TRICHOPHYTOSIS BARBAE.

the parasitic affection, masking in great measure the features of the latter. This was the case in the second of the patients whose case we have cited. Trichophytosis barbae may, moreover, be associated with trichophytosis elsewhere, as in the third case.

PROGNOSIS.—The prognosis of trichophytosis barbae is always good, provided that the physician makes a correct diagnosis, and institutes appropriate measures of treatment. There is no occasion or excuse for the disease to be left unchecked and uncured for months, and even years, as we have known to occur.

TREATMENT.—When left untreated, nature takes care of the affection in the following ways: If the patient be one whose skin reacts but little under irritation, the affection may persist for years, and until the hairs and hair follicles are entirely destroyed. If, on the other hand, there is a prompt reaction with profuse production of pus, this latter may initiate the first step toward a cure by destroying the fungus; for, as Bazin pointed out many years ago, pus is a poison to parasites, at least to the parasite of this disease. This had, in all probability, occurred in the second of the cases reported. When art steps in, she should try to imitate nature's methods to a certain extent; that is, the parasite is to be destroyed as soon as possible. This may most readily be effected by removing the hairs and applying a parasiticide. Of these the more effective are corrosive sublimate, and the oleate of mercury. In our first and second cases Calx Sulphurata was used internally, and in somewhat different doses; the larger dose being given in the papular less acute (plastic) lesion, and the smaller dose in the more acute pustular condition. This is in accordance with the principles we have more fully explained in an article on Calx Sulph. published in the January number of the *Journal of Cutaneous and Venereal Diseases*. In the third case, the affection had lasted for so short a time, and was so favorably located, that internal treatment was unnecessary, the lesion being accompanied with only the slightest evidences of reaction. In special cases various subsidiary measures may be necessary, such as poultices, opening abscesses, etc., which will readily suggest themselves when the occasion arrives.

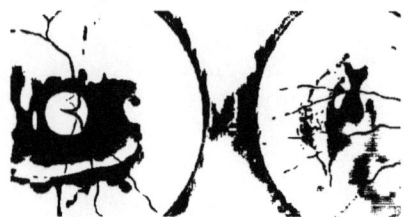

CASES OF RUPTURE OF THE CHOROID.[*]

BY THOMAS R. POOLEY, M.D.,

Permanent Member of the Medical Society of the State of New York.

In my note-book representing ten years of private practice, occur eight cases of rupture of the choroid, from which I select the following six:

CASE I, Fig. 1.—The patient, a man of German birth, aged thirty-four, consulted me January 12, 1875, on account of failing sight in both eyes, but more especially the right, in regard to which he gave the following history. The spring before, he was struck upon it by a missile thrown with great force. He was rendered unconscious for a few minutes by the blow, and for some time after could not see. His sight, however, gradually returned to some extent, and then he noticed that objects appeared crooked and distorted (metamorphopsia).

Soon after the injury, before he consulted me in my office, he was at the Clinic of the Ophthalmic and Aural Institute, and on his card the diagnosis of atrophy of the optic nerves and hemorrhage at the macula was written. The right eye was divergent; its vision $= \frac{20}{200}$. The left, $\frac{20}{40}$; refraction emmetropic. He was blind for green and blue colors.

Ophthalmoscopic examination of the right, showed the appearance delineated in Fig. 1. A curved or crescentic white streak, located between the optic disc and macula, almost in the exact region of the latter, and with its concavity turned toward the disc, about 1½ D. in length, and ¼ D.[+] in breadth, at its widest part. Its edges made sharp irregular lines, and both the edges and middle at about the centre were dotted by black pigment. The small vessels of the retina, which are not numerous in the macula region, passed in an uninterrupted course over the rent. There was not observed, as is often seen, any considerable secondary changes in the choroid. The optic disc in this eye, as well as in the other, showed a moderate amount of atrophy, such as is seen as resulting from the abuse of alcohol and tobacco, to both of which the patient owned up.

In this case we have a fair example of the ophthalmoscopic appearances which rupture of the choroid usually presents; a crescentic-shaped white streak on the temporal side of the optic disc.

The diagnosis of hemorrhage at the macula made at the time of his visit to the Clinic, was quite excusable. That such extravasations of blood occur immediately after the injury to almost any extent, is well known; and later on I shall report a case in which the diagnosis was masked by such

[*] Read before the Medical Society of the State of New York, February 7th, 1883.
[+] D is to signify a unit in ophthalmoscopic measurement, and equals one diameter of the optic disc.

[111]

CASES OF RUPTURE OF THE CHOROID.

an occurrence. The hemorrhage may indeed be enough to find its way into the vitreous, and by causing its turbidity. interfere with the ophthalmoscopic examination.

The occurrence of distorted vision—*metamorphopsia*—as a permanent condition, quite often happens after this form of injury; and, as pointed out by Knapp,* is especially likely to occur after the
sight has temporarily improved, and is due to the contraction of the cicatritial tissue formed in the
choroidal rent.

The retina may be united to it, drawn backward and united to the sclerotic. Then the normal
distribution of the retinal elements, which are of a mosaic-like character, become changed; and their
previous regular retinal meridians are dislocated, so as to produce secondary curves. If they do not
lose their functional power, metamorphop-ia results; but when the sensory elements involved in the
scar are destroyed, a corresponding dark place in the visual field—scotoma—results.

That such defects of vision should occur from such an injury is not, as K. truly says, so remarkable as that, in some cases, vision is but little interfered with.

CASE 2 is one of which I have but very imperfect notes, as I saw it during a visit to Columbus,
and only brought home the sketch with a record of the sight, etc. It is shown in Fig. 2, and occurred in a medical student of twenty-one years, seen March 3, 1877. The rupture is drawn in the in_
verted image. The injury was caused by a blow upon the eye received five years before. The eye
diverged slightly, and sight was very bad, only $\frac{5}{200}$. There was no scotoma nor metamorphopsia.

The ophthalmoscope showed a white, slightly crescentic patch with the concavity turned toward
the optic disc about the same (rather smaller) dimensions as in Case 1, situated above and about
$\frac{1}{2}$ D. from the upper margin (inverted image) of the disc. It lies somewhat obliquely, its inner end
near the nasal side of the disc. There was quite an extensive amount of rarefaction of the choroid
observed, towards the equator of the eye, not shown in the picture; the nerve was white, and the
arteries small—atrophy of the disc.

This case differs from the first mainly in the situation of the rupture, and in the absence of
either scotoma or metamorphopsia. The great loss of acuteness of vision we must attribute to the
atrophy of the nerve, since a rupture of the extent here depicted in such a situation, where the integrity of central vision could be but little impaired, could hardly be held responsible for it. It is
therefore probable that the blow causing the rupture may have in some way (as, for instance, by fracture of the orbit, or hemorrhage therein) compressed the nerve, causing it to undergo atrophic change.

It may be observed that the atrophy of the disc is not very well shown in the drawing. It will
be seen, too, that the loss of vision here was much greater than in Case 1, where the rupture was at the
macula.

CASE 3.—F. G., Fig. 3, a lad of fourteen years, from South Carolina, a student at the St. John's
College, Fordham, consulted me May 31, 1877, on account of the bad vision of his right eye. One year
before, he was struck upon this eye by a tip-cat. He says that the eye was black and blue, and somewhat painful for a few days. He noticed no affection of sight until about one month before his visit
to me, when by accidentally shutting the left eye, he found vision almost entirely wanting in the
upper part of the right. The blindness has gradually encroached upon the rest of the visual field, until

* Archives of Ophthal., Vol. I.

[142]

at the time I examined him, he said he could only see in its outer part. The ophthalmoscope showed, between the macula and disc, two linear white streaks beginning close to the latter, and running in two horizontal, almost parallel branches to the extreme periphery of the fundus; the upper one having a somewhat curved direction with its convexity looking upwards. The lower one had a much straighter course. They were not united by a transverse fissure, but were disconnected at their ends; the upper streak lying closest to the disc, as is shown in the drawing. The streaks were very narrow, not more than $\frac{1}{6}$ or $\frac{1}{8}$ D. wide. Their borders were irregular, and were, especially the upper ones, pigmented. The space included between the rupture showed pigmentation and rarefaction of the choroid. To the nasal side was another rupture running obliquely from above downwards; its upper edge about $\frac{1}{4}$ D. from the disc, 2 D. in length and $\frac{1}{8}$ in width; somewhat linear in shape, the upper end slightly bifurcated, the lower drawn out to a point; along its borders, and in the centre of the gap, a collection of pigment spots. The vessels which could be traced to the locality of the choroidal defects, passed without any change of direction over them.

This case is interesting as showing the occurrence of three distinct rents of the choroid. Those running in a horizontal direction more or less parallel, are an example of the rarest form of choroidal rupture. Of interest, too, is the occurrence of another tear upon the opposite side of the disc. In this case, too, there was at first metamorphopsia, and then a large central scotoma. The sight, as mentioned may be the case, became worse by the contraction of the choroidal scar, causing at first metamorphopsia, and then scotoma; the effect of the contracting cicatritial tissue at first affecting only some of the retinal elements, later all those involved in the scar, and thus destroyed them to such an extent as to abolish their function. A remarkable feature of this case, but not without an analogue, is the insignificance of the symptoms produced by an injury sufficient to cause such serious mischief to the eye. It has been already pointed out by other observers that, although the choroid is a vascular membrane, the amount of hemorrhage caused by its rupture varies greatly. Certainly in this case it would appear that the sight could not have been very seriously impaired, or it would have been noticed soon after the injury. I have seen several cases where the primary results of the injury have been thought very little of.

The next two cases of choroidal rupture observed by me were of the usual linear-shaped white scars situated between the optic disc and macula, and had no features which make it worth the while to report them in detail; they were both examined a long time after the injury which gave rise to them, and in both the vision was considerably impaired. In the one instance $\frac{20}{65}$, in the other, only $\frac{5}{200}$. The latter case was discovered quite accidentally, the patient declaring that he had nothing the matter with his sight. Ophthalmoscopic examination, however, showed a typical rupture of the choroid, and when questioned, he remembered being struck on the eye some years ago with a snowball.

CASE 6, Fig. 4, is the only one which I have had the opportunity to observe soon after the injury. It is, too, by far the most extensive choroidal rupture I have seen myself, or know of as described. The patient was a boy of fourteen years, referred to me by Dr. Arango, his family physician, who was called to see him soon after the injury, and who was so good as to place the care of the case entirely in my hands. The day before I saw him he was struck upon the left eye, as was supposed, by a stone thrown from a sling, by one of his playmates.

CASES OF RUPTURE OF THE CHOROID.

The force of the blow was sufficient to cause him to fall, and remain for some minutes unconscious. The accident occurred in the Central Park, and he was taken to his home in the upper part of the city. Soon afterward he complained of great pain in the eye, and vomited several times. He could not see at all with the injured eye. When I examined him Nov. 14, there was some swelling of the lids, but no ecchymosis. The eye was intolerant of light, and there was a slight degree of circumcorneal injection. There was a moderate, somewhat irregular dilatation of the pupil, a deep anterior chamber, but no tremulousness of the iris. Oblique illumination discovered the fact that there was a small detachment of the iris from its ciliary margin (iridodyalisis) upward and outward, and that its pupillary margin was ruptured below and outward. The ophthalmoscopic examination, which was difficult on account of the irritability of the eye, showed the refractive media to be clear. Between the optic disc and the macula there was a considerable extravasation of blood into the retina, and also one patch of small size near the lower border of the optic disc. The retina in all its outer part was œdematous. Movements of the hand could be seen only on the temporal side. There was no pain. The diagnosis entered in my book was, Œdema and hemorrhage of retina; probably caused by a *rupture of the choroid*. The patient was confined to bed in a darkened room, six leeches put on the temple, a brisk cathartic given, mercurial ointment rubbed into his eyebrow, and a solution of atropine instilled three times a day.

Nov. 23.—Under this plan of treatment, the blood rapidly absorbed, and five days later, it was possible to see a large white patch in the fundus, with the borders still fringed by hemorrhage. I now made the positive diagnosis of rupture of the choroid, which up to this time had only been conjectural. His eye was, however, still rather irritable, and intolerant of light. While directing him to look strongly downward, in order to better see the extent of the rent in the choroid, a small, dark, roundish object came into view, which, on urging him to look still more downwards, fell on the bed. It was a buck-shot, which had been in this position between the lid and the eye since the date of the injury. Upon examining more carefully, I could see a distinct indentation in the globe, behind the equator, and about midway between the corneal margin and outer canthus. This was undoubtedly the foreign body which had inflicted the injury by striking the globe directly, and by the spasmodic action of the lids immediately following, been retained in this situation. He could now count fingers in the upper and temporal part of field. By the 12th of December, not quite one month after I had first examined him, the very extensive rupture in the choroid could be seen. The eye was now free from irritation, the pupil middle wide, atropine discontinued; with eccentric fixation fingers could be counted at 15'.

From this time on, there was gradual improvement; vision rose to $\frac{20}{100}$; and the extent and character of the choroidal rent became more distinctly visible. On January 3 the drawing of the fundus, Fig. 4, was made; it is drawn as seen in the inverted image.

The drawing was carefully made, and is an exact representation of the ophthalmoscopic picture.

The laceration may, for convenience of description, be described as consisting of two parts, a perpendicular or transverse, and a horizontal one, which are, however, united into one. The transverse rupture is curved, with its concave surface turned towards the disc, and situated in the immediate vicinity of the macula lutea. It is about 4 D. in length, and its mean breadth 1 D.; but its centre blends into the transverse rupture.

[141]

Its shape is very irregular, forming three notches above, and two processes or a single bifurcation below; besides this, there is still another notch or prolongation close to the transverse tear.

This transverse rent is continuous with a very much larger one running horizontally, and extending almost to the periphery of the fundus, where it ends abruptly. It is 6 D. in length, and in its widest part 4 D. in breadth. Near the transverse tear it is narrower, but gradually widens out until about the middle of its course, when it again tapers out to become still narrower. The edges are irregular in outline, widely separated, and both the inner parts and margin covered with dark choroidal pigment.

The more central part of the surface is of a whitish-yellow, or gray color, while that nearer the margin is white. All the region above and below the rent, as well as between it and the disc, were strewn with irregular accumulations of very dark choroidal pigment heaps, while on the nasal side of the disc, the choroid looks quite normal.

Some fine blood-vessels can be traced almost up to the margin of the transverse tear, but none were distinctly seen to traverse it.

By the most careful examination in both methods no blood-vessels could be seen upon the horizontal rupture, except on the upper margin near its termination, where a single vessel runs for a short distance along it, and then on to the white surface of the rupture. This vessel could not be traced any farther, nor brought into connection with any other. From the behavior of this blood-vessel and the wide gaping of the choroid, as well as its irregular borders, it may be inferred that the retina is also torn, but of this I could not make myself certain. Whether the retina is involved in the rent will perhaps become more apparent later, by the formation of new connective tissue at its margin; but for the present I must declare myself unable to speak positively on this point. The external appearances of the eye are now nearly normal; the small iridodyalisis being covered by the lid, and the tear in the pupillary margin of the iris hardly visible except by minute inspection. The dilatation of the pupil, and the paralysis of accommodation which also existed, have both disappeared; the pupil is now of about the same size as the other. There is a central scotoma, and a large defect in the field of vision corresponding to the very extensive lesion. Tension of the eyeball is normal, and it is free from pain or injection.

This is by far the most extensive laceration of the choroid I have seen, and if I may judge of those reported by others from their descriptions and drawings, the largest hitherto described.

Although the eye is now free from irritation or pain, it is questionable whether such an extensive injury may not finally result in phthisis bulbi, or be followed by attacks of irritation, redness, and pain from trifling causes.

This case illustrates the difficulty which surrounds making a diagnosis as to the exact nature of the injury soon after its infliction. It is only by taking into account the clinical experiences of others that one can feel moderately sure that the extravasations of blood and opacity of the retina are consecutive to the choroidal rupture. Strictly speaking, the diagnosis of rupture of the choroid is not made until it is seen with the ophthalmoscope, any more than the diagnosis of stone in the bladder is made until the surgeon actually touches it with the point of his sound. There can be no room for doubt that, before the peculiar image which this lesion presents with the ophthalmoscope was first described by Von Graefe in 1854, many such cases were recorded as traumatic irido-choroiditis, hemorrhage into the retina, and so on.

That the blow which gave rise to such an extensive laceration must have been very great in force is shown by the rupture of the iris both at its attached and ciliary borders, as well as by the mydriasis and paralysis of accommodation, to say nothing of the temporarily grave constitutional disturbance produced by it.

In regard to the treatment, I may be allowed to say that the comparatively good recovery in this case may, perhaps, be in a measure attributed to the active treatment employed. Be that as it may, there can be no doubt in any unprejudiced mind as to the propriety of such treatment, when the result of the blow is apt to be followed by grave inflammatory symptoms in the iris ciliary body, vitreous and choroid.

Knapp, in his paper already referred to, says very positively: "Wherever such inflammatory changes, from whatever cause, have occurred, we deem it our duty to institute a serious treatment, of the usefulness of which every unprejudiced ophthalmic surgeon is satisfied. Why should it be inefficient or superfluous in cases where traumatic irido-choroiditis is complicated by rupture of the choroid?" Knapp indeed reports one case which terminated in complete recovery.

It would be going beyond the limits of this paper to enter into any discussion as to the theory of the mechanism of these choroidal ruptures. If any of my hearers are not already familiar with the views advanced, I would refer them to Knapp's article in which he discusses this subject.

I will only say that in this case, the view entertained by Knapp, that the rupture takes place at a point opposite, or distant from the point of injury by *contre coup*, finds an exception, since the transverse part of the solution of continuity was near the point of injury, and might more properly be called direct or *immediate*, than by *contre coup*. I am, however, far from denying that this injury does occur as he suggests. May it not be, however, that when the force of impact from a small foreign body is directly on the globe, the injury is direct, and when the force of blow is transmitted, so as to overcome the elasticity of the membrane, it is received upon the bony surrounding of the eye, rather than directly upon the globe itself.

The retention so long in the cul-de-sac of the agent which caused the injury remaining undiscovered until by accident it came to light, gave me no little chagrin. With that wisdom which comes from hind-thought rather than fore-thought, I now think that the absence of certain symptoms, when I first saw the case, should have led me to infer that the missile had struck the globe, and not the eyelids or surroundings of the eye. Had the latter been the case, there would almost certainly have been considerable swelling and discoloration of the eye; for we all know how easily a black and blue eye is induced, even by an inconsiderable violence.

If I had taken this fact into consideration, I should very likely have sought for and found the foreign body, which, however, did no harm by its residence for so long a time between the lid and eyeball.

SYPHILITIC STENOSIS OF THE LARYNX: TRACHEOTOMY.

BY CHARLES H. KNIGHT, M.D.,

Assistant Surgeon, Manhattan Eye and Ear Hospital.

The patient, a boy sixteen years of age, was brought to the N. Y. Eye and Ear Infirmary, June 19, 1880, suffering with intense dyspnœa and completely aphonic. His face was deeply cyanosed, and wore a frightened expression, and the lad was so exhausted as to be hardly able to walk. It was learned that breathing had been difficult for nearly a week, and that the condition became alarming the day previous. A hurried examination showed excessive deformity of the pharynx and larynx, due to cicatrices and submucous infiltration. The obstruction being evidently laryngeal, and the symptoms urgent, tracheotomy was at once performed, no anæsthetic being used, to the immediate relief of the patient's distress. Nothing unusual occurred in the course of the operation.

On investigating the history of the case it was found that there had been, two years previous, a double interstitial keratitis, and a second attack six months later, resulting in impaired vision, and that three years previous there had been extensive chronic ulceration of the pharynx. A similar, but less severe, attack of dyspnœa occurred about two months before coming to the Infirmary. That attack yielded to mixed treatment, which was so energetic as to cause salivation and loss of most of the teeth.

It was also discovered that the mother had acquired syphilis, possibly from her husband, several years prior to the birth of this boy (her fourth child), three other children having died in infancy of unknown causes; the father is reported to have died eight years ago, of "lung disease," after an illness of more than ten years.

A laryngoscopic examination of the patient two days after the operation revealed entire absence of the epiglottis, and general thickening of the mucous membrane of the vestibule of the larynx, without ulceration, the vocal cords not being visible. The parts were but slightly congested. The pharynx was found to be excessively deformed by cicatricial contractions; the soft palate was adherent to the posterior pharyngeal wall, and the cavities of the pharynx and naso-pharynx communicated near the median line by an aperture barely large enough to admit the tip of the forefinger. There were no points of ulceration. No other indications of syphilis were detected, except those resulting from the affection of the eyes eighteen months previous.

Treatment with mercurial inunctions and iodide of potash internally was attended by subsidence of the laryngeal infiltration to a degree permitting the withdrawal of the canula in two weeks. The tracheal wound healed within a week later.

At that time the voice could not be raised above a whisper, and respiration was somewhat stridulous. Treatment was pursued with gradual improvement until March 1881, a period of nine months, when the patient passed from observation.

The patient again came to my notice at the Manhattan Eye and Ear Hospital in January 1883, when the condition seen in the accompanying plates was found.

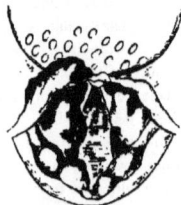

Fig. 141.

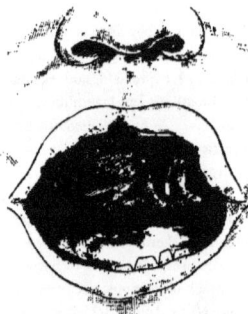

Fig. 142.

The voice is whispering and nasal, resembling that of a person with cleft palate. The vocal cords are not visible, except for a very limited extent at their anterior extremities. Phonation seems to be largely performed by the ventricular bands, which are excessively thickened. There is no apparent loss of tissue in the larynx. The act of swallowing is now accomplished with ease. Up to six months ago, however, liquids would occasionally regurgitate through the nose, or trickle into the larynx.

In general condition there has been great improvement, and there has been no recurrence of laryngeal obstruction. Specific treatment has been entirely suspended since March 1881, no symptoms of syphilis having developed.

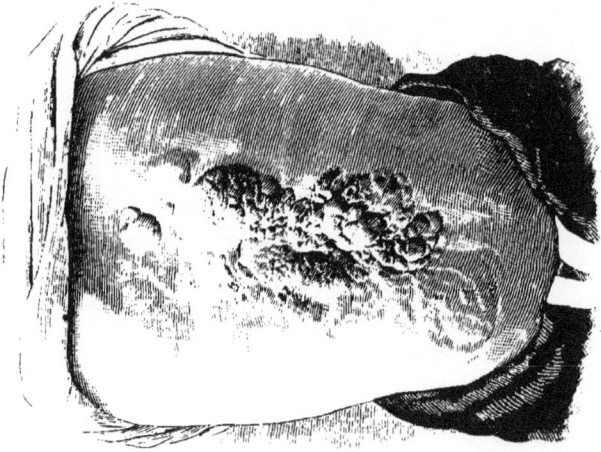

XIV. MULTIPLE SARCOMA OF THE SKIN.

(Cast of Dr. F. Townsend.)

HISTORY OF CASE OF

MULTIPLE SARCOMA OF SKIN.

BY FRANKLIN TOWNSEND, A.M., M.D.

Professor of Physiology, Albany Medical College.

Mr. X., an Irishman by birth, for many years a resident of this country. Married, and a man of family; healthy children; is fifty years of age, and a grocer by occupation.

Family History.—His family history is good; his parents living to a late period, the father dying of pleurisy, the mother of influenza.

Past History.—No venereal taint, except that patient many years ago had an attack of gonorrhœa; from this he recovered after the usual time. He had always been in perfect health until the year 1848, when he had an attack of "inflammatory rheumatism" of both ankles and hips; had also a great deal of pain about his heart at this time, for which he was "cupped and blistered frequently." His illness remained with him "off and on," as he expresses it, for nearly a whole year. Marrying in 1854, he continued well till 1858, when he was a second time afflicted with rheumatism (acute) of both ankles; this attack lasting but a short time. Five years later, or in 1863, patient suffered from symptoms due to malarial poisoning (fever and ague), from which he was not relieved for nearly four weeks, after which time he remained perfectly well, so far as he knows, for about ten years. In 1873 he was "down again," suffering considerably from "internal piles," which were promptly operated upon with entire success.

In the spring of 1874, patient's attention was called to a "breaking out upon his stomach," especially over the right side, "pimple-like in form," and in no marked degree. In from six to eight months the eruption spread to the left side of abdomen. Shortly after this his back became involved by a similar process of "breaking out" upon the skin. This eruption he describes as never itching, and also as being free from soreness. His habits at this time were not perfect, as he frequently drank of malt liquors freely.

From the beginning of the growth to date, there has been a gradual increase in its size. More healthy skin has become involved in the process, while each separate tumor has enlarged in all directions. At no time, patient states, has there been a discharge from the morbid parts, though he believes a change has taken place as regards their color and consistence, viz.: they have become darker, and softer in hue. Mr. X. states also that he never noticed any glandular enlargements about his body until November last (1882), when, after a severe fall upon the ice, he was astonished at feeling a swelling under his right arm, in the axilla; small at first, gradually increasing and becoming more painful with time. Three weeks later, he appreciated a similar swelling in left axillary region. This was much smaller, and not at all painful; besides, it did not grow rapidly.

Present History.—Patient's general health is apparently excellent. No specific conditions observable. Vegetative functions perfect. Drinks of malt liquors moderately and systematically; smokes some, but not excessively. Feels absolutely well, though he thinks that his physical vigor has somewhat failed within " the past few years." The following is the result of an examination of patient's urine as passed for twenty-four hours on Jan. 29th and 30th, 1883.

Chemical Examination.	*Microscopical Examination.*
Quantity passed in 24 hours, 40 oz.	Urates, amorphous, and of soda.
Specific gravity, 1022.	Casts, none.
Color, dark yellow.	Epithelium, from bladder.
Odor, strong.	Pus, none.
Reaction, acid.	Blood, none.
Albumen, none.	
Sugar, none.	
Bile, none.	

The present appearance of the growth, in Mr. X.'s case, is nicely represented in the accompanying photographs. It will be noticed that the nodules and tubercles covering patient's back are of various sizes and shapes, beside being discrete and pigmented. Some are smooth and quite elastic, others firm and hard; a few seem to fluctuate, but no fluid can be withdrawn by the hypodermic exploring needle. Pressure produces no markedly painful impressions. There is no hemorrhage from their surfaces, nor are they ulcerated and broken down at any point. The same might be said with reference to the morbid processes covering his abdomen, with these exceptions, perhaps: 1st. The nodules are smaller and smoother: 2d. Are less discrete: 3d. Are not so markedly colored: 4th. Disease is decidedly less advanced.

The axillary glandular enlargements which were spoken of when giving patient's history, are at the present time of writing very striking, especially those in the *right* axillary space; the left being tolerably free. Under the right arm, in the axilla, is a glandular tumor quite as large as the adult fist, hard, nodular, and freely movable, being in nowise adherent to the overlying skin. It is not tender when pressed, and simply annoys patient because of its size. The left axillary space contains an enlarged gland, possibly the size of a hen's egg; also freely movable, disconnected with the skin, and free from pain or tenderness. The case is interesting, naturally, from the fact that the disease is of rather infrequent occurrence.

PLASTIC OPERATIONS FOR

DEFORMITY OF LOWER EYELID.

BY T. T. SABINE, M.D.

Professor of Anatomy, College of Physicians and Surgeons, N. Y.

Cornelius B. Lawrence, aged fifty. Admitted to St. Luke's Hospital, April 16, 1878. Service of Dr. T. T. Sabine.

The deformity, for the relief of which the operations were done, was the result of syphilitic necrosis of the nasal and right superior maxillary bones, which had been thrown off or removed by operation. The parts about the right orbit presented the following appearance, Fig. 145.

FIG. 145.   FIG. 146.

The line of junction of lower lid and cheek has been drawn backward by cicatricial adhesion in such a way that what is normally the anterior surface of lid is directed downward and backward, the posterior surface with its much chemosed conjunctiva looking forward. By this retraction of the lower lid the upper lid has been so much drawn down, especially at its outer part, that the cornea is almost concealed, and vision is lost for that eye.

OPERATION, May 3d. An incision A B (Fig. 145) was made *down to the bone.* The lid and upper part of cheek were then separated from deep parts by a knife cutting beneath *upper half* of AB toward nose, until the lid could be pushed forward to nearly its natural position. The lifting forward of the lid left a pocket which had to be filled, or the lid would have sunk back to its former mal-position. To obviate this, the following operation was done. An incision was made from B to C *down to the bone*; another from A to C through *skin only*. The skin was carefully detached from the triangle ABC. The flap ABC was then separated from parts beneath, as far as its base AC. It was raw on its anterior and posterior surfaces. Two sutures, armed with a needle at each end, were then passed through the flap near the apex B. The four needles were then successively passed behind the loosened lid (behind the *upper half* of AB) and brought out at E. Traction made upon the sutures dragged the flap behind the raised lid, so that the apex B was brought to E, the flap being concealed by lower eyelid and holding it forward. The sutures were then tied over a piece of elastic catheter.

The space ABC, though but little diminished in size by the traction on the flap, which was dragged through the *upper half* of AB, could have been closed by bringing the edges together, but as this would have made tension on the lid and on the flap beneath it, the following operation was done. An incision was made from C to D, and the flap BCD detached from parts beneath. C was then brought to A, CB to AB, and CD to AD. Puckering along AD was nearly prevented by placing sutures at mid-point.

May 25th. A fold of redundant skin near inner canthus removed.

June 3d. Operation of May 3d not entirely successful. Repeated in nearly same way. Discharged July 11th.

Sept. 19, 1878. Re-admitted. The upper eyelid still drops at its outer part, though not nearly so much as before operation. The lower lid is still too long, so that it is not applied closely to eyeball. To remedy these defects, the following operation (Fig. 147) was done. 1 X, 1 Y are margins of upper and lower eyelids. Triangle 1 2 3 first marked out with knife, and skin removed. The same was done for 1 4 5, and a small part of lower tarsal cartilage at outer canthus removed. The flap 2 1 4 was then dissected up to a *very small* extent at its apex (shaded part), and slid up into 2 3 1, 1 2 being stitched to 3 2, and the *upper half* of 1 4 to 3 1. This raised the outer canthus and upper eyelid to proper position. 5 4 was then stitched to the lower half of 1 4, which remained below the raised outer canthus. This brought the lower lid into its proper place. Discharged cured, Sept. 23, 1878.

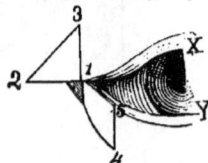

Fig. 147.

XXV. REMOVAL OF EPITHELIOMA BY PLASTIC OPERATION.
(Case of Dr. Alfred Cramer.)

PLASTIC OPERATIONS ON THE FACE.*

BY ALFRED C. POST, M.D., LL.D.,

Emeritus Professor of Clinical Surgery and President of the Faculty in the Medical Department of the University of the City of New York, etc.

CASE I.—*Epithelioma of Face—Excision—Plastic Operation to fill up the Chasm made by the Removal of the Tumor.*

Michael Kerns, aged 61; laborer; admitted into Presbyterian Hospital May 13, 1880; no history of cancer in the family; no history of venereal disease. The patient's habits had been temperate, and he had always enjoyed good health. Three years before his admission, a small pimple had appeared on his left cheek, about three cm. below the lower lid, and subsequently became covered with a dry scab, which frequently became detached. The new growth increased very gradually in size, without pain and without discharge. A year and a half before his admission the vision of the left eye became impaired, and after the lapse of a few months was almost entirely lost. At that time the morbid growth had reached the size of a silver quarter-dollar, without projecting much beyond the level of the integument. Since that time the swelling had increased more rapidly both in circumference and in depth, giving rise to an offensive discharge, and being the seat of a dull pain.

On admission to the hospital there was found a fungous growth of an irregular circular form, about 55 mm. in diameter, occupying the principal part of the left cheek, extending above to within 6 mm. of the lower lid, and below to a line midway between the ala of the nose and the angle of the mouth. The surface of the tumor presented marked irregularities, some portions being of a dull red color, and others of a purplish livid hue. There was no enlargement of lymphatic glands, and no sign of cancerous cachexia.

OPERATION.—On May 15th, the patient having been etherized, after the administration of a tablespoonful of brandy by the mouth and 8 minims of Magendie's solution of sulphate of morphia by hypodermic injection, I commenced the operation by a horizontal incision, separating the morbid growth from the lower lid; I then made another horizontal incision below the tumor, and connected the two by vertical incisions before and behind. I then dissected up, from the adjacent parts, the diseased mass included between these four incisions, leaving a quadrilateral chasm, nearly of a square form. The two vertical incisions were then extended downward, the anterior one to a little below the base of the lower jaw, and the posterior one nearly as low as the base of the jaw. From the inferior

* Read before the American Medical Association at Richmond, Va., May 4, 1881.

[153]

extremities of these two vertical incisions, curved incisions with their concavities looking backward and upward were made to the extent of 5 cm., including between them the peduncle of skin and subjacent tissues for the nourishment of the extensive flap. In the course of these incisions the angular, labial, and transverse facial arteries were divided and secured by ligatures. The flap included between the incisions was dissected up from the subjacent parts, so that its upper extremity could be drawn

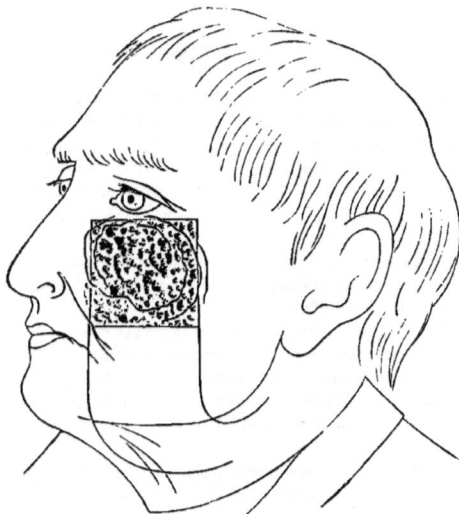

without tension to meet the incision immediately below the lower lid. The parts were then freely washed with a solution of carbolic acid 1 to 40. Before attaching the flap by sutures in its new situation, the ligatures were jerked off from the ends of the divided vessels, and there was no hemorrhage at the time of their removal. The flap was then attached by numerous fine silk sutures throughout its whole circumference, beginning at the upper extremity by which it was connected with the lower lid, and continuing down to the sides of the peduncle. No undue tension was found at any point of the margin of the flap. The length of the anterior margin of the flap in a vertical line was 11 cm.; the breadth of the upper margin of the flap was 55 mm.

Fig. 150.

Fig. 148 exhibits the quadrilateral chasm made by the excision of the tumor and the outlines of the flap with a curved peduncle designed to fill the vacant space.

Fig. 149 shows the flap secured by sutures in its new situation.

8 P.M.—The patient has recovered from the ether. There has been a free hemorrhage under the flap, distending the cheek and the side of the neck to such an extent that several of the sutures have given way, and there is oozing of blood from the open portion of the flap. The house-surgeon made an attempt to arrest the hemorrhage by cold applications, but without success. He then opened the wound, turned out the clots, and secured the bleeding vessels by ligature. The wound was then again washed with carbolic acid, 1 to 40, and was closed as before, by fine sutures, a small drainage-tube being inserted at the most dependent part.

[154]

20th.—The drainage-tube was removed three days after the operation. There is very little discharge from the wound.

30th.—A number of the sutures were removed four days after the operation. The remaining ones have been removed at intervals since. Even those removed to-day were quite dry, having given rise absolutely to no perceptible irritation. There is well marked ectropion of the lower lid.

June 4th.—The wound gapes at its superior border to the extent of 4 to 5 mm.; otherwise it is completely united. There is marked facial paralysis on the affected side. The small granulating surface at the upper border is dressed with ung. lap. calamin.

22d.—The upper edge of the wound is now substantially united, the portion not cicatrized not exceeding 1 mm. in width. The lower lid is fixed in an everted position, exposing the conjunctival surface to the extent of 8 or 9 mm. The junction of the flap with the lower lip is nearly on a level, there being a slight depression. At the junction with the upper lip the depression is deeper. The junction of the posterior margin of the flap with the integument in the temporo-maxillary region is on a

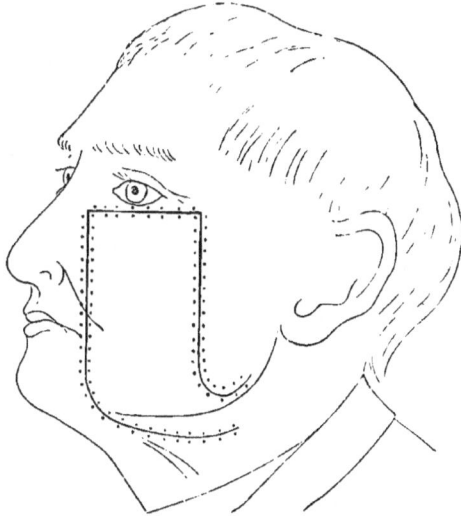

FIG. 151.

perfect level, and the line of union is almost imperceptible. At the junction of the posterior margin of the flap with the integument near the angle of the jaw there is a considerable depression, and the portion of integument between the peduncle of the flap behind and the broader portion of the flap in front is elevated so as to form a prominent ridge about 12 mm. in height. There are two small spaces at the anterior and posterior parts of the peduncle, which are not perfectly cicatrized, and where the granulating surface is covered with a scab.

23d.—I performed an operation for the relief of the ectropion. Brandy, morphia, and ether having been administered, as before the first operation, I detached the eyelid from the cheek just above the line of the narrow cicatrix resulting from the former operation, and then excised the cicatrix from the cheek. I then dissected the integument of the eyelid from the subjacent parts until the ectropion was entirely overcome, and the tarsal margin of the lower lid was brought into close contact with the corresponding margin of the upper lid. The vacant space left between the lid thus dissected up and the upper margin of the flap upon the cheek was about 15 mm. in breadth in the middle and tapering toward each extremity. The whole length of the vacant space was about 6 to 7 cm. The temple being shaved, a flap of sufficient width to cover the space was cut from the integument of the temple and of the forehead, about the outer part of the brow. The inner half of the flap was bounded by straight horizontal lines, and the outer half by curved lines extending upward and forward upon the forehead. In making the flap, the temporal artery was exposed, tied in two places, and divided between the ligatures. Several smaller vessels were also tied. The flap, being loosened from its subcutaneous attachments, was readily brought around to fill the space which had been made by dissecting the lower lid from the cheek. It was secured in its new position by numerous fine sutures. The integument at the junction of the forehead and temple was necessarily thrown into a puckered fold, and a vacant space was left upon the temple behind the peduncle of the flap. This space was dressed with shreds of lint moistened with collodion. At the close of the operation, the patient was troubled with vomiting, but was otherwise in good condition. The newly constructed eyelid presented a nearly normal appearance, except that the flap which had been transplanted had a somewhat livid color.

29th.—The tip of the flap appears to be sloughing.

30th.—Removed the sutures. A large portion of the flap has a sloughy appearance. The edges of the sloughing tissues were dressed with carbolized oil ter in die.

July 7th.—The slough has separated to-day, and is found to be more superficial than it had been supposed to be. The surface was touched with nitrate of silver, and dressed with ung. lap. calamin.

8th.—Collodion dressing reapplied to granulating surface in left temporal region.

17th.—Collodion dressing substituted for ung. lap. calamin. on eyelid.

25th.—The former granulating surface below the left eyelid is completely cicatrized.

August 1st.—Ung. lap. calamin. applied to the raw surface in the temporal region.

28th.—The process of cicatrization in the temple is very slow, there still being a granulating surface as large as a silver quarter-dollar.

September 15th.—The patient was allowed to go out on a pass, but did not return.

In reviewing this case, there are several circumstances which appear to be worthy of special notice.

1st. The manner of preparing the patient for the operation, by administering brandy by the mouth, morphia by hypodermic injection, and ether by inhalation. This is a modification of the practice of Prof. Nussbaum of Munich, and I am in the habit of employing it in all operations which are likely to be protracted. It seems to me to have the advantage of sustaining the heart's action, and of maintaining a more equable and persistent anæsthesia than can be secured by the inhalation of ether without the use of the brandy and morphia. The method is peculiarly adapted to operations upon the face, where the continued application of the ether sponge is embarrassing to the operator, and prevents the operation from being speedily completed.

2d. The large extent of the chasm produced by the excision of the morbid growth and consequently of the flap required to fill the vacant space. It would have been extremely difficult to have transplanted, by the Indian method, a flap of sufficient extent to cover so large a space in this region of the face, and if it had been found practicable, it would have left a very large space to granulate and cicatrize in the region from which the flap had been taken.

3d. The method, adopted in this case, of gliding a flap with a curved peduncle, so as to fill the vacant space without any considerable degree of tension, and without leaving any large surface to granulate. I have operated by this method in a number of cases, and have been able to bring the edges of the wound together by sutures throughout their whole extent and to secure substantial union by the first intention.

4th. The secondary hemorrhage, which rendered it necessary to reopen the wound to secure the bleeding vessels. This was an unpleasant occurrence, although it did not unfavorably affect the ultimate result of the operation. I have no doubt that the hemorrhage was the result of jerking off the ligatures from the vessels, and I would not recommend the repetition of this manœuvre in a similar case.

The result of the operation, although imperfect, may be fairly considered as a marked success.

When we take into consideration the great extent of the disease, and the large size of the flap which was required to fill up the space which was left vacant by the removal of the tumor, we can hardly feel surprised at the occurrence of the ectropion, or regard it as essentially detracting from the success of the operation.

The second operation, which was performed on June 28d, was only partially successful in curing the ectropion. The imperfect success of this operation was the result of partial sloughing of the flap. But notwithstanding this accident, the operation was followed by a very marked improvement in the appearance of the patient, and a decided diminution in the degree of the ectropion. It is a matter of regret that the unexpected elopement of the patient from the hospital not only prevented me from performing an additional operation for the further improvement of the patient, but also deprived me of the opportunity of obtaining a drawing exhibiting the improvement which followed the second operation.

DOUBLE CONGENITAL DISLOCATION OF THE HIP.

BY HENRY B. SANDS, M.D.,

Professor of the Practice of Surgery in the College of Physicians and Surgeons, New York City, etc.

The patient, a girl, aged 14, was admitted into Bellevue Hospital in June, 1872. She had always been deformed, and had occasionally suffered from pain in the right knee during locomotion. Usually she could walk rapidly, and without difficulty; although she had a waddling gait, and exhibited the characteristic rolling motion of the pelvis. Examination showed that on each side, the head of the femur was dislocated backward and upward on the dorsum ilii. The movements of flexion, extension, and rotation were free. Traction on the lower limbs increased their length by drawing the heads of the femora toward the acetabula; but the displacement returned as soon as the force was withheld. No treatment was advised.

The photographs illustrate the characteristic deformity. 1. Each trochanter is seen to be unusually prominent, and to be situated much nearer the iliac crest than when in its normal position. Owing to their projection, the width of the pelvis appears considerably increased. 2. The centre of motion at the junction of the pelvis with the femur being behind the line of gravity, the pelvis is depressed, or inclined forward. 3. To compensate for the obliquity of the pelvis, the lumbar vertebræ present an antero-posterior curvature, with its convexity forward, constituting lordosis, while another compensatory curve is seen in the dorsal region, having its convexity backward.

In the present case a slight degree of lateral spinal curvature is also observed. This is due to the fact that the lower limbs were of unequal length, one being displaced farther than the other.

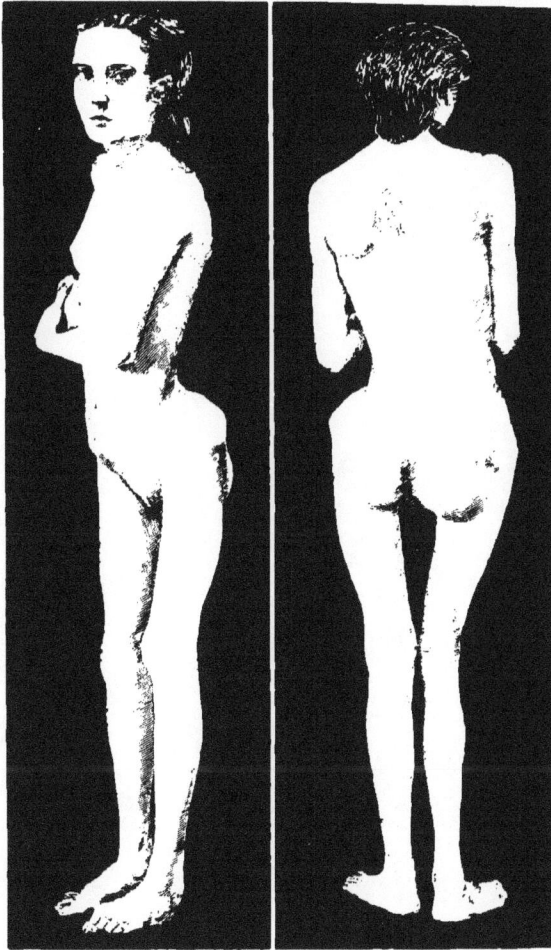

XXVI.   DOUBLE CONGENITAL DISLOCATION OF HIPS.

SARCOMA OF THE ANTERIOR MEDIASTINUM.

BY CHARLES F. BEVAN, M.D.,

Professor of Anatomy and Genito-Urinary Surgery, College of Physicians and Surgeons, Baltimore, Md.

CASE.—W, H., female, aged (19) nineteen, a native of Bremen, came to this country about six (6) years ago, and was at that time in robust, vigorous health. The affection which ultimately produced death, was of about three years' duration. At the time her first symptoms appeared she was engaged in school teaching in Washington, D. C. She first noticed a short, dry cough, void of expectoration, together with pain, at times moderately acute, but generally rather oppressive, which was located always at the same point, viz., the junction of the third (3d) costal cartilage with the sternum on the right side. Domestic medication was resorted to without avail. The affection slowly increasing while her strength gradually diminished, she reluctantly resigned her duties as a teacher in June of 1880, from which time to her death in April of 1882, she remained an invalid generally confined to the house, and under various forms of treatment from the profession and others. She moved to Baltimore, and came under my charge January 16, 1882. Her condition at that time was shockingly bad as regards her nutrition; her form was thin and emaciated, suggesting the decay of chronic tuberculosis. She complained of a cough, most troublesome at night, and attended with very little expectoration; of occasional attacks of dyspnœa, associated with the sensation of oppression and great pain on breathing; of extreme weakness; loss of appetite; of constipation, with attacks of diarrhœa; and that her menstrual flow had not appeared for over five (5) months. Her kidneys acted well; urine normal. Upon examination I found that an oval swelling, three and a half inches (3½ inches) by two and three-quarters (2¾) inches, occupied the sternal region, beginning at the first rib on the right side, and extending obliquely downward and to the left as far as the fourth rib. This swelling was moderately firm, immovable, while the integument glided easily over its surface; it was not painful on pressure, though if handled a great deal, the doing so always gave rise to the old cough; no inflammatory indications were ever associated with it. The mass had grown slowly, commencing in June, 1880. It appeared to be a part of the sternum, or rather a bulging forward of the sternum at the points named, since the sides of the tumor seemed to fade gradually into the breast-bone laterally above and below. A gland as large as a walnut was found above the mamma between the sternal growth and the axilla of the left side; both axillæ were crowded with large glands, which at times were painful. The breast never became involved. In the subclavian triangle of each side, though larger on the right, lymphatic glands were found; one was also met with in the superior carotid triangle of right side. These glands had appeared at various periods between June of 1880 and January, 1882. Both lungs gave normal sounds by percussion and auscultation along the dorsal region, except at the base of the lungs, where slightly diminished resonance was encountered. Over the pectoral region of the right side a dull percussion sound could be elicited, corresponding above to the second intercostal space, two inches to the right of the centre of the sternum, extending obliquely downward as far as the area of cardiac dullness. Sounds of the left side normal. Cardiac dullness enlarged; apex located one inch beyond the mammary line; its sounds were good, excepting an anæmic murmur. Pulse 90. Liver normal; splenic dullness extended as low down as the twelfth rib, but slightly exceeded the axillary lines laterally. The case progressed

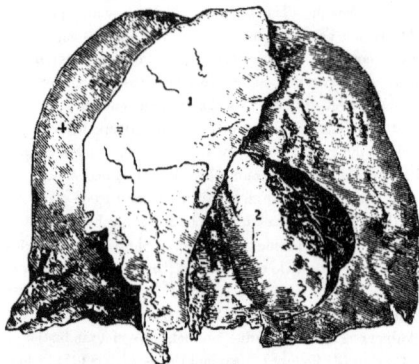

Fig. 154.

from bad to worse during January and February, at which time, February 17th, Prof. O. J. Coskery kindly saw the case with me, concurring in the diagnosis of malignant disease of the anterior mediastinum. In March a general dropsy and anasarca occurred; at one time the abdominal fluid accumulated to such an extent that tapping was contemplated solely with the view of assisting respiration. Under the influence of steam-baths and a diarrhœa, which at this time took place, the effusion rapidly disappeared; the limbs, however, continued very large. A number of needle-punctures were made in them, which largely aided in the removal of the fluid. Unfortunately, however, an attack of erysipelas, confining itself to the punctures of the left thigh, came on, and, with the attendant exhaustion, terminated the case. From March 1st to April 12th the urine was highly albuminous, and contained many hyaline and vitreous casts. An investigation of the family history, made with difficulty, gave the following points. It was learned that the patient's maternal aunt had died from a mammary cancer; another aunt, also upon the mother's side, had died from uterine cancer. A brother of the patient had died within the year from a tumor involving the temporal and occipital bones and the brain, pronounced cancerous by the physician who made the post mortem upon him. A sister died from some disease of the stomach, said to have been cancer, though no post mortem had been made; another sister died from tuberculosis.

Post mortem was made three (3) hours after death, Dr. Coskery assisting. The costal cartilages were cut upon both sides and the sternum removed; strong adhesions existed between the pericardium and pleuræ to the under surface of the sternum. Both lungs, with heart in the pericardium, were removed *en masse*. A large and firm growth measuring three (3) inches at its base or upper part and three and one-half (3½) inches in length was found. This growth was attached to and involved the root of the right lung, surrounding in part the right bronchus; it extended over the base of the heart, arch of the aorta, and pulmonary vessels, so as to effectually hide the aortic branches. Both lungs were smaller by far than normal, though both were thoroughly permeable to air, and neither contained deposits. The amount of pulmonary collapse seemed due to the great feebleness of the patient, which prevented anything like a thorough expansion of the lungs, for no organic changes were noticeable, except that part of the right lung to which the tumor had grown, and at the point for the space of an inch or so the pulmonary tissue was thickened and not permeable to air. Liver was of a dark bronze color, with no secondary deposits in it. Spleen slightly enlarged, friable, and of dark color. Stomach empty, contracted, and without secondary involvement. Bowels normal. A large mesenteric gland was found near the pancreas, which broke down in removing it; its character was cheesy. Uterus, ovaries, and bladder presented nothing of an abnormal type. The kidneys, however, were very large, measuring $6\frac{1}{2}$ inches in length by $3\frac{1}{2}$ in breadth; the capsule was non-adherent, or at least very slightly so;

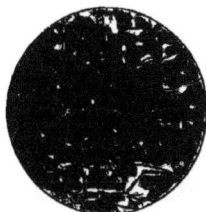

FIG. 155.

in color a pinkish white; on section thoroughly anæmic around the thickened cortical part, while the cones by contrast at least appeared redder than usual. The microscopic character of the large white kidney was unmistakable. Sections of the tumor show it to be a sarcoma of the round-celled variety, though a number of spindle-cells are scattered throughout the stroma.

CANCER OF BREAST—ULCERATION AND DISSEMINATED NODULES.

BY GEORGE HENRY FOX, M.D.,

Physician to the New York Skin and Cancer Hospital.

The accompanying plate is presented as an illustration of cancerous ulceration with disseminated, hemispherical nodules, some of which are undergoing central atrophy. The patient, who was brought to my clinic at the College of Physicians and Surgeons, was 48 years of age, and gave the following brief history. Two years previously the disease began as a small, hard, rounded lump in the left breast above the nipple. An enlarged gland in the axilla soon followed. These tumors were injected by a physician a year and a half later, and shortly after ulcerated. New nodules now appeared with great rapidity on the breast and arm. Many of these ulcerated and coalesced until the axilla became one large irregular ulcer, as is seen in the plate. The borders of this ulcer were hard and raised, and its base presented a dull red glazed appearance. The disseminated nodules were indurated and freely movable with the skin. They were located chiefly in the vicinity of the axillary ulcer, but a few were scattered over the arm. In this case the nipple was unaffected. The disease was very painful, but unfortunately beyond the stage in which permanent relief might be afforded by surgical measures. The patient died about three months after the drawing was made.

XXVII. CANCER OF BREAST.
(*Case of Dr. Gill, Fox.*)

MALFORMATIONS OF THE EXTREMITIES, ETC.

BY J. H. POOLEY, M.D.,

Professor of Surgery in the Toledo Medical College.

CASE I.—Grace S. attracted my attention at first by the absence of the thumb both on the right and left hand, and upon further examination I found other peculiarities worthy of attention, and which I am about to describe.

The child is ten years of age, rather pretty, with dark hair and eyes, tall, and with the exceptions to be mentioned very well developed.

The right arm presents no peculiarity above the elbow, but below that joint seems rather short, and presents a slightly curved appearance, the concavity of the curve looking inward, or to the radial aspect of the forearm.

The forearm measures from the olecranon to the styloid process of the ulna 5½ inches; its motions are all circumscribed and imperfect, from what cause, however, does not clearly appear. The muscles on the dorsal aspect, to wit, the extensors of the hand and fingers, seem to be deficient in power and development, the result of which is a form of partial paralysis, presenting great similarity to the wrist drop of lead palsy.

The hand has four taper, well-formed fingers, but no thumb, or metacarpal bone corresponding thereto; there is a slight contraction of the forefinger.

The left upper extremity is curiously deformed throughout, presenting rather the appearance at first sight of the flipper of a seal or turtle, than of that perfection of grace and mechanical adaptation, the human arm and hand; the following is a detailed description of its peculiarities.

The scapula is smaller than its fellow of the opposite side, narrower, and more sharply triangu-

FIG. 158.

[163]

lar in form. At the humeral extremity it has only one process, which seems to be an exaggerated coracoid, and projects very strongly, giving the shoulder a strangely pointed appearance. This want of the usual rounded contour of the shoulder is still farther increased from the fact that the head of the humerus is small and almost imperceptible either to the sight or touch.

The humerus is short and slightly curved; it measures from the top of the coracoid process to the olecranon 6½ inches.

The left forearm has no radius; the ulna is short, only 4½ inches long, and strongly curved; there are only one or two imperfect carpal bones, those on the radial side being all wanting; and owing to the unsupported condition of the hand, it is strongly adducted.

This hand, like the other, wants the thumb and its metacarpal bone; the fore and middle fingers are contracted, the forefinger overriding its neighbor.

The child's ears are remarkably small, and differ considerably from the usual conformation; she is also quite hard of hearing. She has well-marked phthisis; her mother is also consumptive, and not expected to live long.

The child's mother, I was told by her grandmother, whom she was visiting at the time of my seeing her, presents a similar deformity, except that in her case there is on the radial side of each hand a little teat-like process or appendage, probably rudimentary thumbs. This deformity in the mother, the grandmother refers to a fright she received while pregnant from some crabs crawling about the kitchen-floor, and thinks she is justified in this piece of philosophy, because, as she says, her daughter's hands look like crabs' claws.

Grace had a little sister, who only lived a few months, and was similarly deformed. The other children, two in number, presented no deformity of any kind; they died in infancy. The malformation in Grace and her sister is also attributed to a fright their mother received while pregnant, but in them there is no attempt to trace any analogy between the object producing it and the resulting deformity, but the grandmother says that in both cases the mother confidently predicted at the time of the fright and afterward that the children would be marked or deformed. I have thought this case unusual and interesting enough, with the accompanying illustration, to deserve a permanent record within the reach of any future compiler, or any philosopher who may be glad of one additional instance by which to support or illustrate a theory. It is only a brick, at the service of any builder who has a place into which it will fit.

For a case somewhat similar to this, and an excellent article on the subject, I would refer the studious reader to "A Contribution to the Science of Teratology," by Henry R. Silvester, in the *Medico-Chirurgical Transactions,* Vol. 41, p. 73.

CASE II.—The following case is scarcely worthy of notice or record from the severity or rarity of any one of the abnormalities which it presents; but as a singular and uncommon combination of defects and malformations in the same child, it will merit a place among published cases. November 25, 1869, Mrs. S—— was delivered, after a somewhat tedious labor (face presentation), of a small female child, which presented the following anomalies:

The left leg was short, one inch shorter than the other, and strongly curved. On the left foot there was a supernumerary great toe, which was completely joined by integument and other soft tissues to the natural toe; two distinct phalangeal bones could be plainly felt: the nail was single and large, and divided longitudinally by a deep sulcus or depression.

This double or malformed toe was articulated to the inner side of the metatarsal bone, and stood off at a right angle from the side of the foot. This, the only detail of the case which admitted of illustration, was photographed, and from this the accompanying figure is taken. In all the numerous figures in Annandale on the malformations of fingers and toes, I cannot find one which exactly corresponds with it. On the outer surface of the little finger of the same side, near its phalangeo-metacarpal articulation, was a small tubercle, evidently a rudimentary supernumerary finger.

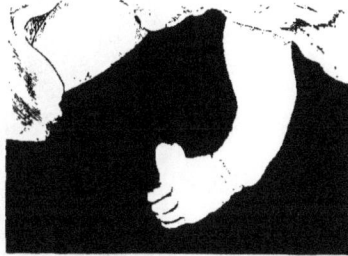

Fig. 159

The little finger of the opposite side deviated from its normal direction in such a manner as to override the finger next to it.

There was incomplete hare-lip, consisting of a fissure of the mucous membrane only, and a rather deep notch on the free border.

The bony palate was uneven, the two sides having apparently come together at a different level; the soft palate was completely fissured.

The tongue was split or bifid at its end, the fissure extending about one inch in depth, and had, moreover, a large gap or deficiency of substance upon its left side about the middle.

The dental arch of the upper jaw was divided on each side by a transverse fissure, so that the tooth pulps, instead of forming a continuous even row, were crowded or agglomerated, as it were, into four distinct heaps. In the lower jaw the alveolar ridge was wanting in front in the region of the incisor teeth, and the place of the gum and tooth pulps was occupied by a series or row of pointed papillar eminences not unlike those found upon the tongue of the larger carnivora. This last deformity gave a very marked and unpleasant depression of the chin, and a hideous expression to the whole face.

Add to all this that the cranium was irregular in form, being much larger and more prominent on one side than the other, and we have an assemblage of deformations, both of excess and deficiency, not often met with in one case.

* This child died December 28th. It never thrived, but progressively emaciated from the time of birth, until at its death it was the merest skeleton imaginable, though no reason could be given for this marasmus, as it sucked well, and never vomited either its mother's milk or cod-liver oil, which it took regularly and constantly. Its bowels never moved until within a few days of its death except from the effect of medicine, though they then acted freely. An autopsy was positively refused, which I regretted very much, fully expecting that it would have revealed the presence of some internal malformation.

I present also with this article an illustration, without any accompanying history, of a museum. It represents the face and head only of a native of British India. Whether this latter had any tumors

Fig. 160.

on any other portion of his body I do not know,—indeed I know nothing about him except that he has one of the most extraordinary physiognomies ever seen among men, hideous enough to serve for a picture of one of his native idols.

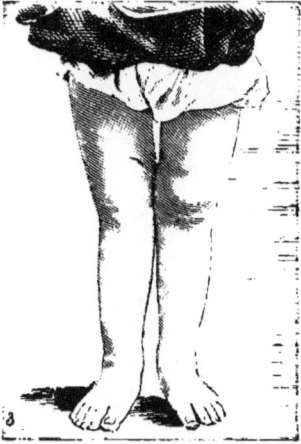

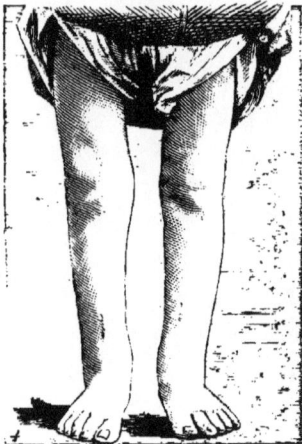

XXVIII THE OPERATIVE TREATMENT OF BOW-LEG AND KNOCK-KNEE.
(Case of Dr. CHARLES T. POORE.)

OSTEOTOMY FOR GENU VALGUM AND CURVATURES OF THE TIBIA AND FIBULA.

BY CHARLES T. POORE, M. D.,

Surgeon to Saint Mary's Free Hospital for Children, New York.

Writers on surgery from the earliest time have advocated bone section or fracture for the relief of deformities, but it does not seem to have been put in practice until the beginning of the present century. Lemercier, in 1815, made a section of the tibia for a fracture of that bone united at an angle; Clémot, in 1834, introduced cuneiform osteotomy for angular curvature; in 1835, Rhea Barton made a section of the femur just below the trochanter minor for anchylosis of the hip-joint at a right angle to the pelvis, and this operation was repeated by both American and Continental surgeons. Langenbeck, in 1852, divided bones with a narrow saw through a small wound, and gave to the operation the name of subcutaneous osteotomy. He operated upon rachitic curvatures of the tibia in this way. Billroth, in 1872, made use of a chisel to divide the bone, operating through a small wound. Since that date subcutaneous osteotomy has become a well recognized operation in surgery. The following paper is intended to be practical, and treats of only osteotomy for knock-knee and tibial curves, and is based on the author's experience.

Tibial Curves.—The vast majority, if not all cases, of curvatures of the bones of the legs occurring in children are rachitic. The deformity is more common among the poorer class living in cities. Yet rachitic curvature of the bone is not unfrequently seen in the children of the wealthy and in those living in the country. In rickets, when the disease is at its height, the bones are soft and easily bent and twisted. The elements characteristic of healthy bone are present, but there is an imperfect formation of the calcareous skeleton of the bone, and in its place there is an excess of medullary tissue and uncalcified bone corpuscles.

As long as an infant is insufficiently supplied with, or is incapable of assimilating, elements necessary for the calcification of cartilage-cells, it is impossible that healthy bone can be produced. But as soon as the power of assimilation is regained, and the proper food is provided, the cells rapidly abstract the lime-salts, and soon become more dense and harder than normal bone. It is evident that no rule can be laid down as to the age at which a rickety child's bones will become sclerosed; it is not a question of years, but of nutrition. I have met with very hard bone in a rachitic child three years old, and I have seen very soft bone in a boy of ten years. Again, the hardening process may be much more advanced in one limb than in the corresponding one; thus one tibia may be very hard, the other quite soft. The immediate cause of all rachitic curvatures is mechanical, either using the limbs while the bones are soft, or the manner in which the nurse carries the child, bend the weakened bones, and deformity is the result.

Curvatures of the bones of the legs may be either in a lateral or anterior direction; they may involve the whole bone from epiphysis to epiphysis, or the deformity may be confined to either extremity. Curvatures of the lower end of the tibia are generally of early origin, and the angle is apt to be very acute; the bone may be twisted on itself so that the spine points more or less inward or outward. In anterior curves, not complicated with lateral ones, the whole bone is more generally bent. In some cases, when the bending is low down, the tibia overhangs the foot, giving a very peculiar appearance to the limb. Sometimes the tibia is flattened laterally and the spine is very sharp and thin.

There are certain deformities of the legs due to improperly treated fracture, to inflammation of the bone, and to repair after limited necrosis.

Genu Valgum.—Without going fully into a discussion of the cause and pathology of knock-knee, it may be stated that this deformity, coming on in children under eight years of age, is rachitic, and is due to changes taking place at articular ends of the bones forming the knee-joint, by which the internal condyle assumes either relatively or actually, a lower plane in its relations to the external condyle than in the normal limb; or similar changes occur in the relations of the heads of the tibia. There may be a true hypertrophy of the internal condyle of the femur; an atrophy or flattening of the external condyle; the shaft of the femur may be bent at its lower third convexity inward, thus depressing its internal articulating surface; the inner head of the tibia may be increased in height, or the external head depressed; the epiphysis of the femur or tibia may have become twisted; or two or more of these conditions may contribute to make up the deformity. By these changes in the ends of the bones the long axis of the tibia is thrown outward, and genu valgum is produced.

Genu valgum first making its appearance between the fifteenth and twentieth year is not rachitic. It may make its appearance in apparently strong persons whose height has increased very rapidly and who have been subjected to fatiguing labor, as standing on the feet all day at some heavy work, jumping on one limb from a cart many times a day, etc.

It would seem to be dependent on a yielding state of the ends of the bone due to the fact that growth has been so rapid that the young bone-cells have not been able to assume the hardness, and stability of mature bone, and that, while in this state, undue and uneven pressure, beyond its sustaining power, had been placed upon it. In all cases that I have examined at this period of life the deformity had taken place at the epiphysis. The condition is sometimes accompanied by tenderness along and a little beyond the cartilage. I have also seen some effusion into the joint.

While this process is going on in the bones, and before any great amount of sclerosis is present, correction can be made by the use of appropriate braces; but after the bones have become hard, the deformity cannot be cured except by operative measures, and for this osteotomy and osteoclasis are generally performed.

Osteotomy.—In order to prevent repetition, the mode of performing an osteotomy is here given. There are two forms of bone section, the linear and cuneiform. For a linear osteotomy an osteotome is used. It resembles a chisel gradually sloping down on both sides, like a knife-blade, to a sharp cutting edge. They should be made of the best steel, and well tempered. It is well to have three of the same width, but of different thickness, and one much narrower for section of the fibula. The chisels are made similar to a carpenter's chisel. A sand-pillow, a knife, and heavy mallet complete the

necessary outfit. Having rendered the limb bloodless, an incision is made at the point of desired section, immediately down to the bone in a line parallel with the long axis of the limb of sufficient
length to easily admit the osteotome. On the knife as a guide, the osteotome is passed down to the
bone and then rotated so as to be at a right angle with the line of the incision. It is then to be driven
into the bone by pretty firm blows with a mallet. After each blow the instrument is to be rotated
on its long axes, so that it will not become wedged, and for the purpose of dividing the bone in different
directions. I do not use any spray or take any further antiseptic precautions, except to keep the wound
wet with a carbolized solution. After sufficient of the bone has been divided, the osteotome is removed,
and the fracture completed by bending the bone. A sponge dampened with carbolized water is placed
over the wound and held in place with one or two turns of a bandage. The Esmarch bandage is then
removed, and the other limb operated upon in the same manner. After completing the division of
the other limb, the sponge is removed from the leg first operated upon, and with a sharp pair of scissors any pieces of fat or cellular tissue that may protrude from between the lips of the wound are
removed. After which a long piece of adhesive plaster, about two-thirds of the width of the wound,
is placed transversely across it, so as to bring its edges into perfect coaptation. The uncovered portions on either side will allow an exit for any undue accumulation of blood; a small compress is then
placed over the wound. The whole limb is bandaged with a flannel bandage some distance above the
point of section; if a leg, above the knee, and over this a plaster-of-Paris splint. Before the latter
dries, the limb is to be put into a proper position. On the third day a fenestra is cut, about three
inches square, over the seat of the wound, and the latter examined; if all looks well, no further care
need be taken of it. If, however, there is any discharge, a fresh compress is to be put over it, and
renewed every day or two. At the end of four weeks the splint may be removed.

There will be enough oozing from the wound to discolor the plaster-of-Paris splint more or less.
The object aimed at in this care of the wound is to secure primary union, and thus reduce the chances
of any complication. It will be found, on examining the wound on the third day after a linear osteotomy, if all has gone well, that the cut will be represented by a mere line, that there is no swelling or
redness about the line of incision, and that all oozing has ceased. Indeed, it is often difficult to see
just where the cut had been made.

Genu Valgum.—The operation generally adopted for this deformity is Macewen's; the section
is made on the shaft of the femur, above the line of the epiphysis. The leg being strongly flexed and
the thigh rotated outward, an incision is made down on to the ridge running from the tubercle for the
attachment of the tendon of the adductor magnus to the linea aspera, at a point corresponding to a
line drawn a finger's breadth above the superior tip of the external condyle; this is above the epiphysial
cartilage. Section is commenced with the largest osteotome, so as to make the cut in the bone as
wedge-shaped as possible. The instrument should be placed on the above-mentioned ridge, and driven
backward and outward, its direction being changed after each blow, in order to divide the bone in all
directions. As soon as the osteotome becomes wedged, it is to be withdrawn, and the next smaller one
substituted in its place, to be again replaced by the smallest, if necessary. After the bone has been
divided on its anterior, inner, and posterior aspect, the osteotome is withdrawn, and fracture completed
by grasping the femur just above the point of section with one hand, and the leg with the other, and

carrying the latter inward. After division the wound is to be treated as pointed out above. In correcting the deformity, after the plaster-of-Paris bandage has been applied, care should be taken that the lower fragment does not slip backward. I have seen this in one of my cases, and although it did not interfere with the ultimate result, yet it is well to be avoided.

In some few cases, when the deformity is very great, it is necessary to make a section of the bones of the legs; this can be done after consolidation has taken place. The time required to obtain firm union is about four weeks.

Curvatures of the bones of the legs may be treated by osteotomy or osteoclasis. Those cases where the curves are long, involving the middle portion of the limb, if there is not much anterior curvature, are best treated by osteoclasis. Deformity just above the ankle or below the knee are not adapted for this operation, because there is not sufficient room to apply an osteoclast without endangering the epiphysis.

Osteotomy for tibial curves is of two kinds, the linear and cuneiform; the former is applicable for the correction of lateral and slight anterior curves; the latter for anterior curves alone. In operating upon the bones of the legs the fibula should *always* be divided, and the section should be made *first*, because, after fracture of the tibia, it is difficult to steady the smaller bone. I usually divide the fibula in the direction of upward and inward, so that the fragments may slip by one another.

The tibia is to be divided at the point of greatest curvature, and the incision in the skin is made in the long axis of the limb down on to its crest, and the bone partially divided, and then fractured. The fibula is to be fractured at a point corresponding to that at which section of the tibia was made.

Cuneiform osteotomy is performed through an open wound. An incision is made down upon the crest of the tibia at the point of greatest curvature, of sufficient length to give plenty of room, and the periosteum separated from the bone on its anterior aspect to the extent of the size of the wedge to be removed, which should be accurately determined before commencing the operation. In cutting, the straight edge of the chisel should be kept toward the part of the bone which is to remain in the body. It is also best to remove first a smaller wedge, and then, by successive cuts enlarge it to the required size. The apex of the wedge should extend beyond the central cavity in the bone into the compact tissue; the remaining bone can be divided with an osteotome. After removing all loose pieces of bone, I think it well to make a counter-opening on the inner side of the leg at a point corresponding to the lower limit of the wedge; through this and out of the operation-wound carbolized horse-hair is passed to act as a drain. The hemorrhage is more profuse than in a linear osteotomy, and the accumulation of blood as apt to cause too great tension, and prevent primary union. Any piece of fat or cellular tissue is removed that may protrude from between the lips of the wound. The edges are then accurately united with carbolized gut, the horse-hair being spread out and passed between the lips of the wound; over the wound a compress is applied, and the limb put up in plaster-of-Paris; and before the splint hardens the limb is brought into the desired position. On the following day the horse-hair is removed—piece by piece—the compress reapplied. By adopting this plan the wound will unite without any suppuration or discharge, and consolidation takes place in about four weeks.

The accompanying illustrations are added showing the results of osteotomy.

CASE I.—C. B., 8 years of age, was admitted into St. Mary's Hospital in February, 1882. He had uncomplicated genu valgum; he gives the history of having suffered from rickets when a baby, and at time of examination has the remains of that disease in enlarged epiphyses, a slight antero-lateral curvature of the tibia, and a depression of the external condyle of both femora, which was the cause of the knock-knee. Plate No. I. is from a photograph taken shortly after admission. There is no relaxation of the ligaments about the knee-joint. He appears to be a perfectly healthy child. He has some difficulty in walking owing to the deformity. On March 6th osteotomy was performed on both limbs, the femur being divided just above the epiphysial cartilage, and the limbs put up in plaster-of-Paris. On the 8th, fenestra cut.

On March 15th, patient complains of some pain in right thigh; temperature 103; and on examining the thigh above the point of section, an abscess was discovered extending up between the unscular planes. Wound was reopened, and compress applied over the abscess. By the 20th all signs of pus had disappeared, and four weeks after the date of operation patient was up. Plate II. is from a photograph taken just before he left the hospital. He has been seen since; he walks well; and there has been no return of the deformity.

I report this case because it is one of the few that have been accompanied with an abscess. The pus did not communicate with the bone, and its presence did not seem to retard recovery. I can give no satisfactory reason for its advent.

CASE II.—G. F., age 5 years, was admitted with rachitic curvature of both tibiæ, in both lateral and anterior direction; both lateral and anterior bending are marked. In February, 1882, osteoclasis was performed on both limbs, but after fracture it was found impossible to correct the anterior deformity, although the lateral could be easily straightened. The limbs were put up in plaster-of-Paris. On May 4th, cuneiform osteotomy was done on both tibiæ, the wound drained with horse-hair, and edges of wound brought into perfect coaptation with sutures. Fig. III. is from a photograph at time of operation.

On May 5th, the horse-hair was removed, and on May 11th the wounds were found to be closed; there had not been a particle of suppuration, and her temperature had never been above 99°. At the end of four weeks the splints were removed and patient allowed to go about.

In my two first cases of cuneiform osteotomy suppuration followed the operation. I think it was due to the fact that I did not make a counter-opening and drain properly. Since I have adopted this plan I have experienced no trouble; the wounds have closed like those made in simple osteotomy. Fig. IV. shows the result and also the scars of the incision.

CASE III.—M. P., aged 4 years, had rachitic curvature of both tibiæ. She walks with a rolling motion, and has frequent falls. Fig. V. exhibits the amount of deformity. February 15, 1879, linear osteotomy performed on both limbs, both tibia and fibula being divided. The bones were very hard, so that section was made very slowly. About the middle of March the splints were removed and patient allowed to get up. During the following week there was some tenderness and swelling about the seat of fracture in left leg, and there was a slight yielding; plaster-of-Paris bandage reapplied. In April the splint was removed; union seemed firm. She was discharged in May, walking well. Fig. VI. exhibits the result.

In one other case I have seen some yielding after an osteotomy.

After the patients get up it takes a week or so to regain the use of the limbs. I have never seen any stiffness follow an operation either for genu valgum or bow-legs.

The results of osteotomy, both as regards life and the correction of the deformity, are good. In almost a hundred operations I have never lost a patient from any cause that could be attributable to the operation. I have lost two children from diphtheria; at the time of death the wounds had closed and the bones consolidated. I have had two cases in which an abscess appeared between the

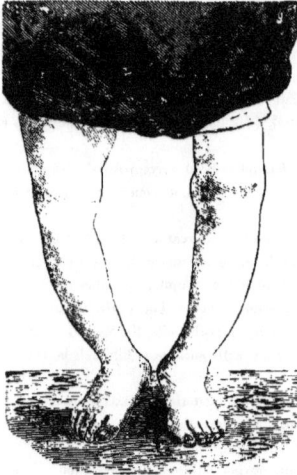 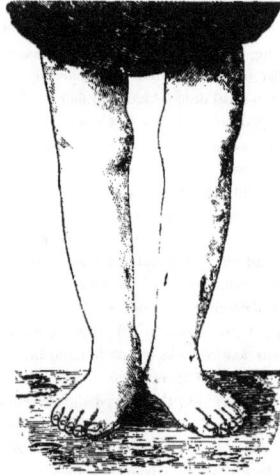

Fig. 165.   Fig. 166.

muscles after an osteotomy for genu valgum; one, my first case, after a linear section for bow-legs, and two in which suppuration followed a cuneiform osteotomy for anterior curvature of the tibia. The two latter cases were due to the fact that I did not manage my wound properly; since then I have experienced no trouble.

There is no shock following these operations. The amount of pain is very slight; often no anodyne is necessary, and as soon as the effects of the ether have passed off, the only complaint generally made is from the confinement.

The cases used as illustrations in this paper are not typical ones in regard to the course of the wound. They are used to show that notwithstanding the presence of an abscess the good results were obtained.

[172]

STILL BIRTH FROM UNUSUAL CAUSE.

BY EDWARD L. PARTRIDGE, M.D.,

Professor of Obstetrics in the New York Post-Graduate Medical School, Attending Physician to the Nursery and Child's Hospital.

Mrs. ——, aged 25, pregnant for the first time, having reached the end of gestation, was taken in labor at 1 A.M., May 17th. Membranes ruptured at this time, and cervical dilatation proceeded in a tedious way, attended, during a period of thirty-eight hours, by unremitting, regular uterine pain. During this stage of dilatation, chloral and morphine secured some needed relief. So severe was the

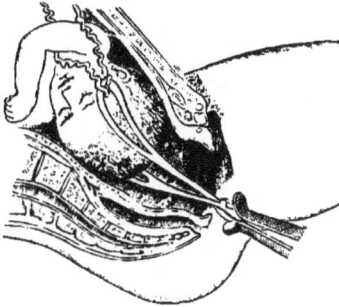 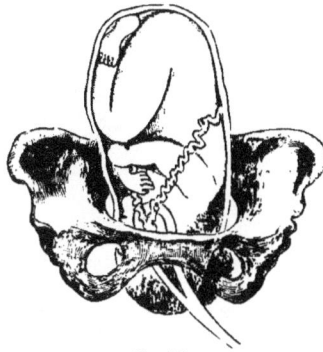

FIG. 167.   FIG. 168.

labor that when I first saw her, at 10 P.M., May 17th, a caput succedaneum was well formed, though the os was less than half dilated.

May 18th, 10 A.M., os was two-thirds dilated, and as she remained in the recumbent position during the day, there occurred as the result considerable œdema of the anterior lip of the cervix. Pulse continued good until 2 P.M., when it became somewhat irritable, and during an hour and a half preceding application of forceps was steadily in the neighborhood of 100.

At 3 P.M., the os being three-quarters dilated, the forceps was used, just before their application the fetal heart being very distinct. There was no difficulty in applying the blades, and three-quarters of an hour were occupied in delivery,—traction, with no compression, being employed. Up to the time that the head passed the cervix, there was no staining of the vaginal discharge by blood. When the head had so descended that the finger in the rectum readily kept it from receding during intervals of contraction, I proceeded to take off the forceps. The lower blade was easily removed, but the upper blade was held at its upper extremity, so that gentle manipulation failed to release it. Thinking that the ear might lie in the fenestra, I preferred to use no force, and with the succeeding pain the occiput emerged so as to reveal a *loop of funis* lying *between* the extremity of the blade, which was still applied, and the right side of the vulva. (See Fig. 169.) It was pulseless, and had been prevented

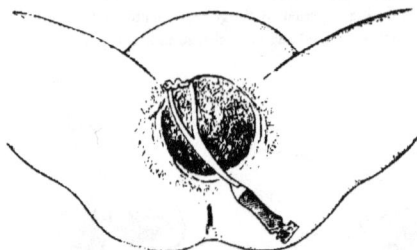

FIG. 169.

from slipping up as the head descended, by being *caught and held by the fenestra of that blade*, lying, as the accompanying wood-cuts show, *not* between the forceps and the head, but between the pelvic wall and the extremity of the blade of the forceps, and having been compressed in a way fatal to the child.

Attention to the child demonstrated its death. All action of the heart was absent, and there was no sign of any reflex activity. The funis, which was above the average length, *was not coiled either around an extremity or about the neck.*

The mother made a good recovery. The case is of interest as illustrating an unusual cause of still birth. If the forceps had not been employed, it is extremely probable that the close apposition between the head and the maternal structures would have prevented any descent of the funis between the head and the pelvis, and fatal compression would not have occurred. The forceps operation was demanded, however, by the circumstances of the case, and there was *no way* by which any knowledge of the position of the funis could be ascertained in this, or in any similar case, by an examination which would be either called for, or even justifiable.

It would be highly improper, of course, as a routine practice, when the forceps was to be used, to push the hand past and above the head, in a uterus from which the liquor amnii had escaped thirty-six hours before, and nothing short of this operation could have demonstrated the relation of the funis to the child.

[174]

APPARATUS FOR FRACTURES OF THE LOWER JAW.

BY J. S. WIGHT, M. D.,

Professor of Operative and Clinical Surgery at the Long Island College Hospital.

In 1877 I began to treat fractures of the lower jaw with the apparatus shown in the accompanying drawing (Fig. 170). Since that time I have treated twelve simple and four compound fractures of the lower jaw with this apparatus, and it has been better than any other apparatus that I have used.

Fig. 170.

I have found this apparatus especially adapted to the treatment of compound fractures of the lower jaw. In using this apparatus the surgeon can readily combine two principles of practice, namely, retention and manipulation. I find that the fragments of a broken lower jaw, as time goes on and repair takes place, may be more and more manipulated and pressed into a proper position by the surgeon. And the manipulation should take place from day to day, and then the fragments of bone will be well retained in place by the apparatus, which may be described as follows:

A quadrangular piece of wire cloth of convenient size, as seen in the figure, is cut, and the meshes on each side are removed, so as to give us open buckles on either side of the head. A quadrangular piece of muslin is now cut, and three tails are made on each end of it, as is shown in the drawing. The three tails of the bandage are applied as follows: The end of the middle tail is put into the middle buckle of the head-piece; the posterior tail of the bandage is put into the anterior buckle of the head-piece; and the anterior tail of the bandage is put into the posterior buckle of the head-piece. The tails of the bandage on the other side of the head are applied in a similar way. It will be seen that there is a pad of moderate thickness on the head under the wire-cloth.

APPARATUS FOR FRACTURES OF THE LOWER JAW.

The sub-mental splint is made of fine wire-cloth, and can be easily understood by reference to *a* and *b* in the drawing; *a* represents a quadrangular piece of wire-cloth cut into a little way at each end, while *b* represents the same piece bent up at the ends and sides for application under the chin. The beard may be used for a pad. When the fracture is compound, with an opening on the outside, oakum makes a good pad for the sub-mental splint. The wire-cloth can be washed every day with a disinfectant solution.

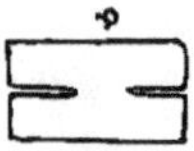

FIG. 171.    FIG. 172.

This apparatus has the following advantages:

It is easy to make and apply; it does not readily get out of place forward, backward, or laterally; it can be made loose or tight, as the case may require; it will permit some change in the direction of the pressure on the sub-mental splint; and it can be easily kept clean.

XXIX. HERPES FAC ALIS
(Case of Dr. G. L. Jackson)

AN UNUSUAL CASE OF HERPES FACIALIS.

BY GEORGE THOMAS JACKSON, M. D.,

Clinical Assistant to the chair of Dermatology, Col. Phys. and Surg., N. Y. ; Assistant Physician to the N. Y. Polyclinic (Dept. Skin Diseases), &c., &c.

On Wednesday, August 1, 1883, Christopher T., twenty-one years of age, unmarried, born in this city, and a case-maker by trade, presented himself at the N. Y. Polyclinic for treatment. He is a strong, healthy-looking young man, not at all of the nervous temperament, and has always enjoyed good general health. Some fifteen months ago he had a genital sore and a bubo, but no secondary eruption.

On Saturday last, July 28th, he was playing ball in the hot sun. He became overheated during the game and was wet by a slight shower. After the game he drank beer to excess while still sweating freely, and became intoxicated. On Sunday, July 29th, he was in bed till three o'clock in the afternoon with nausea and vomiting, chill and fever. He was feverish during Sunday night, and on Monday morning early he had a burning pain in the face, followed by an eruption of vesicles upon the middle of the upper lip. The burning pain continued during Monday, and by evening there was an eruption of vesicles over the whole of the right cheek. During Monday night he had such intense pain in the face that he could not sleep, and on Tuesday morning the left side of the face was covered with vesicles. During Tuesday his fever left him. His bowels are regular. His sleep is disturbed on account of the pain in his face. He has a splitting headache.

The eruption occupies the anterior portion of the face, included between lines drawn perpendicularly through the outer canthus of each eye, or rather a point about one-half inch beyond it. It is composed entirely of unruptured, fully distended vesicles, with cloudy contents, seated upon a reddened base. These are confluent upon the lower part of the cheeks, and upon the lips and chin, and form patches of varying size upon the rest of the affected portion of the face. There are but four patches above the level of the eyes ; three of them over the left eyebrow, and one small patch immediately above the anterior upper portion of the left ear. The vesicles upon the right side of the face seem to be slightly more developed than those on the left. The greater part of the vermilion border of the lower lip is taken up with the remains of broken vesicles. There are no vesicles upon the tongue, the inside of the cheeks, or soft palate, but the pharynx is reddened and congested.

He was directed to smear a little vaseline over the affected portions of his face, and over all to freely dust on a powder composed of equal portions of starch and oxide of zinc. He was cautioned against rupturing the vesicles, and was given a one-grain pill of opium to procure sleep.

[177]

AN UNUSUAL CASE OF HERPES FACIALIS.

August 3d.—Had to take one opium pill on the 1st and 2d inst. to procure rest, as his face burned at night. Patient says he feels a good deal better to day. The vesicles are drying up without breaking. There is a good deal of swelling under the eyes, and a well-marked conjunctivitis of right eye. No new vesicles. He was ordered a saturated solution of boracic acid for his eye.

August 6.—No pain or soreness to-day. The vesicles have desiccated into thick crusts, which on being raised expose a delicate new epithelium with no tendency to ulceration. Directed the patient to remove the crusts carefully after soaking with oil, and then to apply ung. zinci oxid. as a protection. The eye is better.

August 8th.—Patient reported himself as feeling first-rate. Many of the crusts have been removed or fallen off, leaving fresh smooth new skin. Eye entirely well. Ordered a continuance of ung. zinci oxid. Patient did not report again.

This case is the most extensive in its distribution of any case of herpes that I have ever met. The patient had never had "fever blisters" or "cold sores" before. The case which stands next in its area of distribution, as far as I know, is one figured in Fox's Photographic Illustrations of Skin Diseases. It is that of a boy nine years old. In it the eruption was confined to both cheeks, with a small patch on the corner of the mouth. He was subject to repeated attacks of "fever blisters."

REMOVAL OF A CAPILLARY NÆVUS AFTER FORTY-TWO YEARS' STANDING.

BY J. S. WIGHT, M. D.,

Professor of Operative and Clinical Surgery at the Long Island College Hospital.

Mr. H., a carpenter, single, and 42 years of age, was sent to me by Dr. Andrew Otterson in January, 1883. He had on the upper and anterior aspect of his left thigh a large vascular tumor, represented in the accompanying drawing, Fig. 175. The tumor was pendulous, hanging down the front of the limb, and measured about twenty-four inches in circumference, extending down the front of the thigh about seven or eight inches, and having a very large neck. A large sinus distended with blood came down the thigh obliquely inward from the base of the tumor, from which it is seen projecting. A space about three inches in diameter on the lower end of the tumor had ulcerated and was oozing blood from time to time. I advised the patient to submit to an operation, and with this object in view sent him to the Long Island College Hospital.

When the patient was admitted to the hospital, I put him in bed, and gave him tonics and nutrients to prepare him for the operation. In a few days the tumor became considerably smaller, but would enlarge again on the patient's getting up. According to the statement of the patient's aunt, the tumor at birth was a small capillary nævus, about as large as the ordinary finger-nail. His mother declined to have it operated on, and it had been growing for forty-two years.

On the 24th January, 1883, at my college clinic, I removed the tumor and the sinus in the following manner: I procured a strong ligature about ten feet long, and put on it six large curved needles, so that they were equal distances apart, as represented in the accompanying drawing, Fig. 176. The drawing represents the base of the tumor with the ligature in place after the needles have all been pushed

Fig. 175.

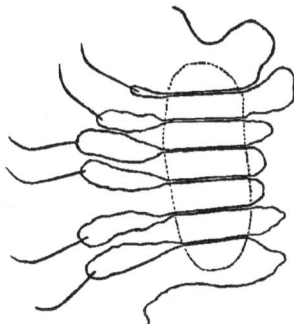

Fig. 176.

through the tissues underneath. The ligature was cut at the eye of each needle, giving seven shorter ligatures, which embraced the base of the tumor in seven parts; while this was being done the tumor was held up by my clinical assistant Dr. Rand. The separate ligatures were then tied in succession with great force, in order to strangulate the base of the tumor. Then the sinus above noted was transfixed at its base and ligatured in the same manner. The base of the sinus was transfixed, in order to keep as far as possible from the femoral vessels. I then cut off the tumor rapidly with my scalpel just outside the ligatures, when a large quantity of blood was discharged from that portion of the tumor cut off, and the stump bled freely from a number of open sinuses, and oozed blood freely from some parts of the cut surface. Several of my artery forceps and general pressure restrained the flow of blood. The sinus that had been ligatured was not cut off. It is worthy of remark that the patient began to take ether and then objected, saying that he would have the operation performed without an anaesthetic. I felt that my patient was in earnest, and directed the discontinuance of the ether. He was perfectly quiet and made no sign of pain during the operation, which occupied about twenty-five minutes.

After the operation the patient had a high temperature—sometimes as high as 105°—and suffered great pain, and was in peril from threatened secondary hemorrhage. In about two weeks the stump of the main tumor, as well as the sinus, came away. This result was greatly facilitated by the application of the elastic ligature and by the final application of strong silk ligatures. At that time the denuded surface measured ten inches in length and five inches in width, and is represented in the accompanying drawing, Fig. 176.

A microscopical examination of the tumor made by my colleague, Prof. E. S. Bunker, exhibited an abundance of sinuses, but no arteries nor veins. The connective-tissue skeleton of the tumor was extensively infiltrated with indifferent cells. And it is possible that at some future time sarcomatous growths may form not only where the tumor was removed, but also in other parts of the body. A definite conclusion on this point cannot now be reached.

A significant clinical fact may be noted, namely: As soon as the stump of the tumor came away the patient did not have any more surgical fever, apparently demonstrating that the absorption of the waste products was the cause of the general febrile disturbance. The blood-vessels had been gradually closed underneath the surface that had been laid bare by the ligatures as they cut their way through the tissues. The granulations appeared to be healthy, though they were somewhat exuberant, but they were kept down by the application of sulphate of copper. On February 26th the granulating surface was contracting and forming scar-tissue around its border, and the patient felt well and was sitting up. On the 1st of April the ulcer was nearly healed, and my patient has the prospect of a complete recovery and left the hospital at his own request in order to go to work. Since operating on this patient I have removed a large nævus from the face of an infant by the same plan of ligation, and the result has been very satisfactory.

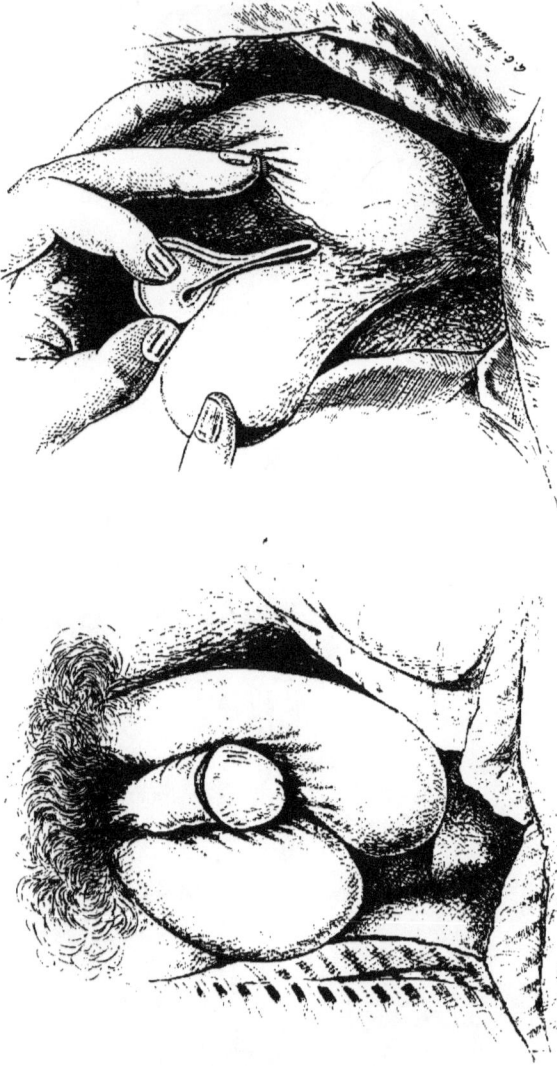

From a sketch from life expressly for Illustrated Medicine and Surgery.

XXX. HYPOSPADIAS SIMULATING HERMAPHRODITISMUS.

(Case of Prof. J. L. Little.)

SPURIOUS HERMAPHRODITISM.

A CASE OF HYPOSPADIAS, WHERE THE PATIENT, MISTAKEN FOR A FEMALE AT BIRTH, HAS PASSED AS SUCH TO THE PRESENT TIME.

BY JAMES L. LITTLE, M.D.,

Professor of Surgery, Medical Department of the University of Vermont ; Professor of Clinical Surgery in the N. Y. Post Graduate Medical School ; Surgeon to St. Luke's and St. Vincent's Hospital, N. Y., &c.

This patient came under my care about eight years ago, with the following history: He was thirty-four years of age. At the time of his birth his mother was attended by one of the most prominent physicians of the town in which she lived, who pronounced the child a girl. Between the age of twelve and fourteen, however, he found by his own observation that he differed from other girls of his acquaintance, and calling his mother's attention to it, she consulted a physician, who, after making an examination, informed her of the nature of the deformity, and assured her that the child was a male. The parents proving to be too ignorant to properly comprehend the difficulty, and notwithstanding the assurance of their physician to the contrary, continued in the belief that the child was a female, and in consequence made no change in his apparel. As a result of this stupidity on the part of his parents and his own modesty and want of courage, together with an amount of religious superstition seldom met with, he has grown to his present age, still wearing the garb of his mistaken identity, and passing as a female among his acquaintances; although he is aware that it is generally whispered about the town by many who know him that he is an hermaphrodite.

He bears a female name, and when he consulted me at my office his appearance was more that of a man than a woman. His form was strong and masculine, height about five feet six inches, and his walk and actions were decidedly those of a male. He wore his hair long and in the style usually affected by females of his position in life, but he also said that he had a beard which he was obliged to shave every morning in order to escape detection.

His occupation is one usually performed by males, and although he is a skilled workman he receives only the compensation usually paid females.

Upon examination I found that he was suffering from the congenital deformity called hypospadias of the third degree. The corpora cavernosa and glans penis seemed to be well developed, with the exception of the meatus. The corpus spongiosum and urethra consisted only of a gutter lined throughout with mucous membrane, commencing at the glans penis and ending at the junction of the penis with the scrotum. The scrotum was cleft or divided into two lobes, each containing a well-developed testicle. In addition to this he had a bubonocele of each side.

Upon lifting the penis between the finger and thumb, as represented in the accompanying cut, which was drawn for me at the time by Mr. Geo. C. Wright, anatomical artist, of this city, it seemed as though a simple division of the cutaneous tissue would be sufficient to permit the straightening of the organ. I recommended that this operation be performed, but told him that it was extremely unlikely that I would be able, owing to the extreme degree of the malformation, to increase the length or efficacy of the deficient urethra. I advised him to enter the hospital, but here a number of difficulties presented themselves. He could not, of course, enter a female ward, being a male; nor, on the other hand, could he be put in the male ward still clad in the garments of the other sex, and these he objected to laying aside, as, he claimed, that he would not like to return to his home, even after an operation, dressed as a man after having passed so many years as a woman. Then too, even though the curvature of his penis were corrected, unless it was found to be possible to so increase the length of the urethra that it should end at or near the point where the meatus should have existed, he would not be able to micturate without assuming a sitting or squatting position; and he declared that the feminine vestments in view of that necessity were more convenient than those of the male. He also felt that he could not then spare the time necessary for the operation, as his aged father and mother were both dependent upon him for support.

His sexual desires were very strong, and were a source of almost constant annoyance to him, as he associated continually with females, but he had never made an attempt at sexual intercourse for fear of exposing his true condition. Some time ago he wrote asking the advisability of castration, with a view to putting an end to his sexual appetite, and he stated at that time that he suffered from lascivious dreams and emissions.

His erections were imperfect, the penis making almost a complete curve, so that the glans pointed almost to the scrotum.

The case was seen and examined at the time of this visit by Drs. T. G. Thomas and T. M. Markoe, of this city.

This case, although not presenting an unusual deformity, is interesting from the fact that the patient has all his life passed as a female. The deformity *itself* is due to an arrest of development of the genital organs during fœtal life. The genital furrow which exists in the fœtus has not entirely closed so as to form the urethra. The development of the corpus spongiosum has also been arrested, and this gives rise to that curvature of the penis which is so difficult to overcome successfully by an operation.

The deformity may exist in three degrees:

First. Where the opening of the urethra is just behind the frenum.

Second. Where it is between the frenum and scrotum.

Third. Where it exists just at the junction of the penis with the scrotum.

In regard to the treatment of these cases the plan to be adopted depends upon the degree of the malformation. Cases of the first and second degree are frequently amenable to operative procedures. The operation consists in the establishment of a urethra in the direction of the natural channel, and to describe its details would require more space than is allowed me for this article. The great difficulty in many cases is to overcome the curvature of the penis, owing to the fact that the fibrous structure covering the corpora cavernosa is non-extensible. Incisions through this will sometimes enable the surgeon to straighten the penis at the time of the operation, but if inflammation sets in, and the corpora cavernosa become involved, their structure is likely to be permanently impaired.

A CASE OF HEREDITARY DEFORMITY.

BY E. P. WILLIAMS, M.D.,

Physician to the New York Dispensary.

Sarah R., aged about ten months, came under my professional care about three years ago. I saw her occasionally for a year or eighteen months, when she passed from my observation. I have since heard that she died of one of the ordinary diseases of childhood. She was somewhat feeble, but not more so than many of her class and surroundings. In nothing would she attract notice except for abnormalities represented by the cuts. The right arm and forearm (Fig. 179) were perfectly formed, and in all respects normal; the wrist joint was natural; the movements of flexion and extension of the hand upon the forearm were perfect. Some of the carpal bones were absent, but which ones could not readily be determined. The metacarpal bones and the phalanges of the ring and little fingers were absent. This hand was a very serviceable member; the absence of two fingers seemed to cause no inconvenience whatever.

The left forearm and hand (Fig. 180) presented the most marked deviation from the normal type. The forearm had but one bone, which possessed some of the features of both the radius and

FIG. 179

FIG. 180.

the ulna; supination of the hand was perfect, but it could not be completely pronated without a movement involving the arm itself. The wrist joint was imperfect, as the movements of flexion and extension of the hand were only about one half as much as normal. The forearm was capable of full extension by the volition of the child, but was generally carried partially flexed upon the arm; the movement of flexion, however, was not complete; no part of the hand could be touched to the head of the humerus, nor could the chin or nose be touched by the finger without a movement of the shoulder joint. That some of the carpal bones were absent was evident, but it was not easy to decide which. The metacarpal bones, with the phalanges of the middle, ring, and index fingers, were all absent. The thumb was freely movable in all directions; it was usually partially flexed, but was extended at the pleasure of the child. In the use of this hand she seemed to suffer but little inconvenience from the loss of three fingers, though it was evidently lacking in strength and in the ability to grasp and hold as large objects as the other hand could. Without including the absent carpal bones, the number of which could not be accurately determined, twenty-one bones were missing from the two extremities.

The mother was a fairly healthy woman, she had borne three children; the first, whom I saw often, was a healthy girl, perfect in all her members. The second child, a girl, whom I never saw, died at about a year old. The mother informed me that she was affected very much as the subject of this sketch, only that in her case it was the right hand which had but a thumb and one finger, and the left hand had thumb and two fingers; just the opposite of her next younger sister.

The father was a man of unusual health and vigor; and at times a traveling showman. His right hand possessed but a thumb and one finger, while it was the left hand of Sarah which was so affected. I had no opportunity of knowing whether there was more than one bone in the forearm. The left hand had the index and middle fingers; the thumb was somewhat distorted, large, bunchy, and not freely movable. This hand was exhibited as a specimen of a "dog's-head hand," to which it bore some fancied resemblance when closed: the finger on the right hand was shown as a "snake finger," to which it bore no resemblance. These defects were not hereditary in his case.

Those who believe that mental impressions of the mother while pregnant may affect the offspring can find a corroboration of their belief in this case, while those who do not so believe may look upon these abnormalities as merely a natural transmission of defects. Mrs. S. informed me that when she was "carrying" the first child she was living with her husband in good circumstances, and had no cause for worry or anxiety; this child was born perfectly developed. During her second and third pregnancies her husband was "on the road" nearly all the time, and she was constantly anxious in regard to him. These two children were born deformed, and partook of the deformities of the father. The parents have lived together almost constantly since the birth of the third child; a fourth child has been born to them which I have not seen, but am told that it is perfectly formed.

Figs A

Figs C

Figs B

MUCOUS TUBERCLES OF THE OS UTERI AND VAGINA.

BY FESSENDEN N. OTIS, M.D.,

Clinical Professor of Venereal Diseases in the College of Physicians and Surgeons, New York City; one of the Visiting Surgeons to Charity Hospital, Blackwell's Island; Consulting Surgeon to St. Elizabeth's Hospital. New York City. etc.

The foregoing sketch represents the appearance of the upper part of the vagina and the os uteri of M. W., a prostitute, who presented for examination with the statement that she was quite free from any venereal disease. The circumstances which led to her examination were as follows: Mr. W. H., a young gentleman of 20 years of age, called complaining of a slight muco-purulent urethral discharge which was first noticed the day previous and about a fortnight subsequent to his first and only sexual connection. The discharge was associated with slight itching at the urethral orifice; slight smarting on urination. The lips of the orifice were quite red and pouting, but no erosion could be discovered in the examination of the first inch of the canal with a urethral speculum. The length of time intervening between the connection and the discovery of the discharge was opposed to its origin from gonorrhœal, leucorrhœal, or menstrual secretions. The probability seemed rather in favor of a syphilitic infection. With the view, as far as possible, of settling this point, the young woman with whom connection was had, was induced to come to my office for examination. She denied all knowledge of any sore about her privates then or previously. Examination of the faucial region showed deep congestion of the fauces and pharynx; no erosions. The lymphatic glands of the cervical, epitrochlear, and inguinal regions were markedly enlarged, but free from tenderness. There was no appearance of any eruption on the body nor any trace of a previously existing eruption. Examination with the speculum revealed the presence of fully half a dozen well-marked mucous papules on the vaginal walls, just within the ostium, and several in the superior portion, and on the cervix, as shown in the colored plate. One was found between the labiæ on either side, and fully a dozen, ranging in size from a flattened peppercorn and a large pea, surrounded the anal orifice. The girl, on being questioned, admitted the knowledge of their presence, and stated that she first noticed them three or four months previously. As, however, they gave her not the slightest pain or inconvenience, she had supposed them to be warts and of no consequence. She was aware of some leucorrhœa, but not more than was habitual, to her. The appearance of these lesions were perfectly characteristic of the mucous tubercles of syphilis.

The young man was not treated, but kept under observation. The urethral discharge gradually disappeared, and in about a fortnight his urethral orifice resumed a healthy appearance. There was slight enlargement of two or three inguinal glands in either groin, which continued for some weeks longer, and finally disappeared. Not the least further trouble indicating a syphilitic infection occurred during the next six years.

Remarks.—The evidences of syphilis in the case of this girl were positive, in the presence of characteristic mucous tubercles and well-marked painless gland enlargements in all the usual localities —most pronounced in the groins. The escape of the young man from an attack of syphilis, while most extraordinary, is in proof of the necessity of a previous fracture of the skin or mucous membrane in order to produce a syphilitic infection. His trouble was apparently the result of the irritating character of the secretion of the mucous papules. but which was not sufficient to effect a lesion sufficient to permit the introduction of the syphilitic influence.

VENEREAL WARTS OF THE PRÆPUTIAL ORIFICE.

BY WALTER L. RANNEY, M.D.,

Assistant to the Chair of the Practice of Medicine in the Medical Department of the University of the City of New York, Attending Physician to the University Medical Dispensary.

J. I. K., age 27, came to my office in January, 1882, giving the following history:

Eight months previously he had contracted a gonorrhœa, for which he did but little in the way of treatment, and the discharge, which was rather profuse, continued for nearly four months. About three months from the commencement of the discharge he noticed a few small growths at the end of the prepuce, and these had gradually increased in number and size until he came to me for treatment. On examination I found the condition pictured in the drawing, which I made at that time. The prepuce was very long, extending nearly half an inch beyond the glans; and growing from and surrounding the orifice hung a mass of vegetations, grouped in pyramidal segments. This mass measured a trifle over an inch and a quarter in its longest diameter and about an inch transversely. By a little manipulation I was able to retract the prepuce completely, the growths then surrounding the corona like a ring. At the meatus there were similar growths, smaller and more vascular, but grouped in the same manner. There was also a slight gleety discharge, which the patient had not noticed. The glans itself was entirely free from these warts, with the exception of one or two very small ones at a short distance from the meatus; and the whole trouble being evidently due to the continuance of the discharge and to uncleanliness, I advised circumcision. The patient declined to undergo any operation, and I treated him at irregular intervals for some weeks with a drachm of corrosive sublimate to an ounce of collodion, applied locally, and with a consequent diminution in size of the growth of over one-half. He then consented to be circumcised, and the discharge having been checked, there has been to the present time no indication of any return of the vegetations at the meatus or on the glans.

[187]

THREE CASES OF FRACTURE OF THE SPINE, WITH AUTOPSIES.

BY L. EMMETT HOLT, A.M., M.D.,

OF NEW YORK.

Late House Surgeon to Bellevue Hospital.

CASE I.—*Fracture of the lumbar spine—paraplegia—complete recovery. Autopsy twenty-two years later.*

Wm. R., aged fifty-nine, was admitted to Bellevue Hospital, October 23, 1880, complaining of sharp neuralgic pains with a feeling of weakness in the lower part of his back. He gave the following history:

Twenty-two years ago, while at work upon a chimney, he fell a distance of forty-two feet, striking upon his back. He was picked up unconscious, and taken to the New York Hospital. He was unconscious for a week, and completely paraplegic for three weeks, the bladder and rectum being affected. In three months he began to walk, and in ten months from the injury was able to resume his work. He had had no previous trouble with his back. The kyphosis was not seen before the accident. Since that time he had led an active life, wearing usually a wide cloth belt about his loins, but suffering from no symptoms referable to the spine until September last, when darting pains in this region began. He tired easily, being able to walk but a few blocks at a time.

On examination he was found to be thin and anæmic, and his general condition was poor. There was a bony prominence in the lumbar region, involving apparently three or four vertebræ, from about the last dorsal to the third lumbar. It was about one-half inch in height, and was not sharp. There was no lateral deviation of the spinous processes here. Firm pressure and percussion elicited some tenderness. He stooped easily and without pain. When he attempted to bend the body backward he suffered very great pain, and could not carry it beyond the erect position. Lateral movements were also difficult and painful. The flexors of the thighs did not seem to possess the usual power, especially those of the right limb. Otherwise no loss of power nor anæsthesia could be discovered on careful examination.

The patient died four months after admission from fibroid phthisis and cardiac disease.

As it does not pertain to the present subject, I will omit an account of the autopsy, except in so far as it relates to the spine. The condition of the cord and the membranes is unfortunately omitted in the account of the autopsy. The last dorsal and the four upper lumbar vertebræ were removed, and these are shown in the accompanying plate. The last dorsal and the third and fourth lumbar vertebræ seemed normal. The first and second lumbar were fused by new bony deposits anteriorly. The bodies were firmly anchylosed, posteriorly they were separated by intervertebral substance. The articular processes of these vertebræ were united by ossific matter, and were somewhat irregular in shape. From the posterior part of the arch of the second lumbar vertebra a slender osteophyte projected to the extent of half an inch. The intervertebral foramina and the spinal canal were not encroached upon. The whole specimen presented a very decided convexity backward, the reverse of the normal spinal curve in this region.

[188]

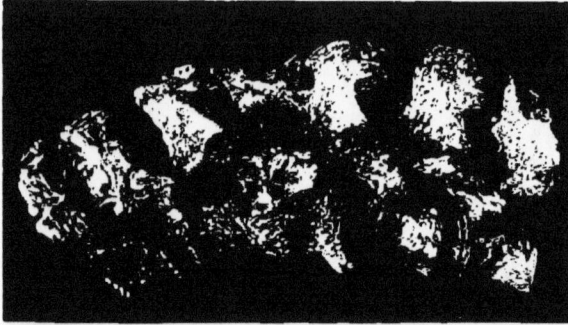

Fig. 184.

Fig. 185.

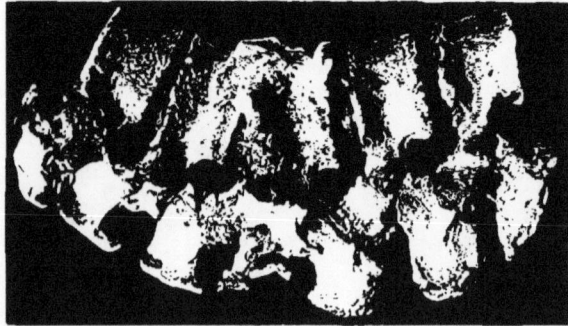

Fig. 186.

XXXII. UNION OF FRACTURED SPINE

(Case of Dr. L. E. Holt.)

THREE CASES OF FRACTURE OF THE SPINE, WITH AUTOPSIES.

My own interpretation of the specimen is as follows: There was an impacted fracture of the anterior part of the bodies of the first and second lumbar vertebræ; secondly, a fracture of the articular processes uniting these bones, and considerable new osseous tissue thrown out; thirdly, the face of impaction caused the projection backward of the vertebræ and the separation between the spinous processes of the two most involved. This separation is of itself an important sign of fracture, even when backward and lateral displacement are both absent, or very slight. My attention was first called to this point by Prof. E. G. Janeway. I have verified it in a number of cases. This case is interesting, as it shows that notwithstanding the severity of the original injury, the complete paralysis and the bony deformity, the man was able to follow for twenty years his trade as a mason.

CASE II.—*Dislocation of the fifth cervical vertebra, complicated with fracture. Attempted reduction under ether followed by complete paraplegia. Death two months later of meningo-myelitis.*

John S., aged fifty-seven years, was admitted to Bellevue Hospital, January 10, 1881. He stated that he had been perfectly well until December 29th, when he fell from the rigging of a ship a distance of about twenty feet, striking, he thought, upon his shoulders. He was stunned for a few minutes, but soon recovered himself, and walked that afternoon a considerable distance, feeling as well as ever, except that he was obliged to carry his head far forward, and was unable to straighten his neck. Stomach, bladder, and rectum were not affected. A few days after this he gradually became aware that his arms, especially the right one, were growing weaker. This increased so that by the end of the week he was unable to feed himself or button his clothes, though he could do both well before. Numbness and tingling in the fingers accompanied these symptoms. He had noticed nothing abnormal in his lower extremities. No pain was felt except in attempting to move the head. This caused darting pains in the arms near the shoulders, which were sometimes very severe. He walked to the hospital. He was a very stout muscular man in excellent general condition. He walked freely; nothing abnormal found on examination of the lower extremities; sensation and motion seemed perfect. He held his head peculiarly; his neck was flexed and the chin was carried far away from the sternum.

Dr. Frank W. Olds, one of the internes of the hospital, made the accompanying drawing (Fig. 187) of the patient at the time, which is an excellent likeness both of the patient and his deformity. He had been obliged to hold his head in this way since the accident. It was found he could rotate and flex the head freely. Any attempt at extension, however, produced severe pains in both arms. The same result followed any trial at passive movements in the same direction. The spinous processes of the seventh cervical and first dorsal vertebræ seemed to be a little more prominent than usual. Firm pressure here caused the referred pains, which were not produced by pressure elsewhere. Concussion of the spine caused no pain. No head symptoms were present; no paralysis of facial or ocular muscles; pupils normal. The pulse was 80; respirations 24 per minute, and regular. There was almost complete paralysis of the right upper extremity. The muscles of the left upper extremity were much stronger, possessing, perhaps, one-third normal power. No anæsthesia could be discovered in either arm, nor in fact anywhere, though careful examination for this was made.

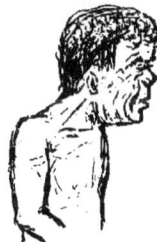

FIG. 187.

The patient was put to bed, and absolute rest enjoined. Two days later, under the advice of

[189]

THREE CASES OF FRACTURE OF THE SPINE, WITH AUTOPSIES.

Dr. Alex. B. Mott, the patient was etherized by Dr. W. B. Vanderpool and myself, with the design of making an attempt to reduce the deformity; the steadily increasing paralysis being thought sufficient to justify a trial of this dangerous expedient. Even under ether the position of the head did not change until some degree of force had been used. This was applied by making gentle traction with one hand under the occiput and the other under the chin. Meanwhile, slight pressure was made upon the projecting spinous processes. After this manœuvre was repeated once or twice cautiously, the head was found freely movable. and the prominence had in great part disappeared. The head was brought into line and firmly held while the patient came out from the ether. When next seen, two hours later, he was found completely paralyzed in all four extremities, with complete anæsthesia in all and in the trunk up to the level of the fourth ribs. The urine was retained. Before going under ether the patient's pulse was 108, temp. 100¼° F.

During the next few days the temperature rose steadily, and once reached 105°, the pulse ranging from 90 to 120, and the respiration about 30 per minute. He complained much at first of some pains about the shoulders, later became delirious; his tongue was brown and cracked, and he vomited nearly everything taken by the mouth. In spite of all precautions bed-sores formed in two weeks. Then the acute febrile symptoms subsided in a measure, and his anæsthesia disappeared almost entirely, and he regained considerable power in the left arm, but not elsewhere. His respiration was diaphragmatic. The greater part of the time he had amusing hallucinations, but no active delirium.

Early in March he had an exacerbation of his fever, the temperature ranging from 102° to 104¾°. He grew more and more feeble, and died March 7th, of pulmonary œdema. He had no convulsions, and was conscious till within an hour of his death, though his mind has been clear very little of the time since the febrile symptoms came on.

Autopsy made by Dr. Wm. H. Welch fourteen hours after death.

Head.—Skull normal ; an abundant exudation of fibrine and pus covered the base of the brain. There was pachymeningitis interna with a good deal of exudation extending up from the meninges of the cord over the whole posterior fossa of the skull. The brain itself showed nothing abnormal.

Chest.—Calcific deposits about the aortic valves, but heart otherwise normal. Old pleuritic adhesions over the right lung, both lungs very markedly œdematous.

Abdomen.—Liver and spleen normal. Slight cloudy swelling of cortex of the kidneys, but nothing more; no pyelitis.

Spine.—The entire length of the cord was the seat of general inflammation of the dura and pia mater with copious purulent exudation. This was more marked in the dorsal and lumbar regions. The exudation was chiefly confined to the posterior surface of the cord. There was in the cervical region considerable peri-pachymeningitis. Macroscopically the cord showed on its surface the evidence of inflammation. It was hardened for microscopical examination, but unfortunately was lost before this was made. The cervical and upper dorsal vertebræ were removed, sawn through vertically, and are shown in the accompanying cuts. (Figs. 188 and 189.)

An anatomical anomaly was found. The bodies of the sixth and seventh cervical vertebræ formed a single piece, without any trace of intervertebral substance. The articular processes between them were also fused. The laminæ and spinous processes were distinct. This seemed to have nothing to do with the injury. except perhaps determining the point of giving way just above. The principal

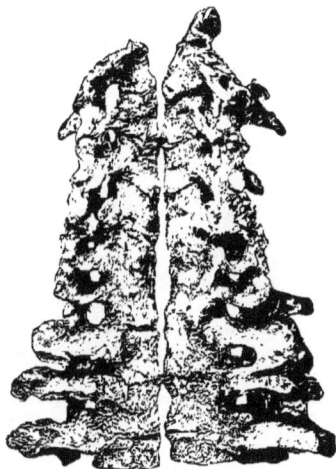 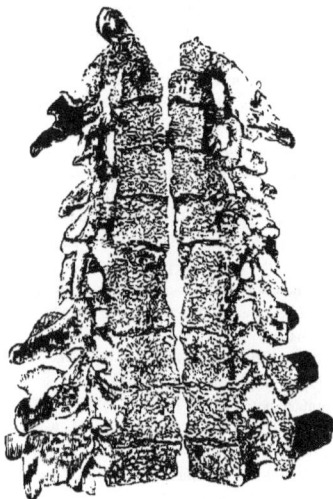

FIG. 188   FIG. 189

lesion was a dislocation forward of the upper five cervical vertebræ to the extent of nearly one-fourth of an inch. This involved a breaking down of the intervertebral disc between the fifth and sixth, and a laceration of both the anterior and posterior common ligaments. The intervertebral substance anteriorly was not torn, but the front part of the body of the sixth vertebra was chipped off downwards and forwards for about half an inch. These small fragments had become quite firmly united by the callus, but the irregular line of fracture could be distinctly traced.

Viewed laterally, the right side showed ruptures of the ligaments uniting the articular processes of the fifth and sixth vertebræ with a separation of the articular surfaces. On the left side the corresponding articular processes had been fractured, but were united by callus with considerable displacement.

Viewed from within, after the bones had been sawn open, the spinal canal was found greatly encroached upon by the displacement backward of the upper part of the body of the sixth vertebra. It was narrowed in its antero-posterior diameter at least one-half. The intervertebral foramina for the sixth pair of cervical nerves were diminished, the left one to a mere slit about one line by three, in size, the right being only a little larger. This was due in considerable part to the callus which had been thrown out. The rest of the foramina were not perceptibly affected.

An interesting question in this case is, what effect the attempt made at reduction had upon the displacement. I cannot avoid the conclusion that, although we overcame an undoubted external

[191]

deformity, we did increase the pressure upon the cord, perhaps by unlocking some slight impaction which had occurred. The gradual paralysis which had been coming on before this seems most likely due to pressure from inflammatory deposits and from callus about the nerves at their exit. That the pressure up to this time had not been much upon the cord itself is shown by the escape of the lower extremities. Our experience in this case will certainly not encourage others to attempt reduction in similar cases.

CASE III.—*Partial dislocation of the sixth cervical vertebra, with fracture of the pedicles and processes, and paralysis of both upper extremities. Death from an accidental complication.*

Peter K., aged thirty-seven years, was admitted to Bellevue Hospital, December 7, 1880. On the morning of admission he was thrown from a wagon-seat to the pavement, striking, as he thought, on his back and left shoulder. He did not lose consciousness, but was unable to rise. He was an intemperate man, but denied that he had been drinking at the time. When seen by the ambulance surgeon, an hour later, there was paralysis of both upper extremities, the left being more marked than the right. Here there was complete drop-wrist. There had been no vomiting. He was able to stand, but staggered if he attempted to walk. He was put into bed, and the following morning I made a more careful examination. The bowels and bladder had acted normally up to that time. The left upper extremity was found almost completely paralyzed. The shoulder muscles, however, seemed unaffected. He could flex the forearm very feebly, and just move the thumb and index finger. There was no other voluntary power in the member. Almost complete anæsthesia existed below the elbow, and sensation in the arm was greatly diminished. He could execute all the usual movements with the right arm, but with only about one-fourth the normal power. The anæsthesia was distributed as upon the opposite side. Sensibility in the trunk, face, and lower extremities was normal. There was very little, if any, loss of power in the lower extremities. He could walk with a little assistance, but with a tendency to stagger and fall like an ataxic patient. There was no facial paralysis. He protruded the tongue normally. There was no dyspnœa; deglutition normal. Examination of the spine, both externally and by the pharynx, failed to reveal any deformity. A point of tenderness, on pressure, was found over the spinous processes of about the fourth and fifth cervical vertebræ. He was ordered to be kept perfectly quiet in bed for the present.

Two days later the patient became wildly delirious during the night, and fell out of bed. The following morning he was quieter, and in fact was quite rational. During the day he grew worse again, and had to be tied in bed. At last he became so violent as to necessitate his removal to the cells, where he remained until his death. His morning and evening temperature had been 100° F.; his pulse ranging from 110 to 135. Full doses of bromide of potassium and chloral were ordered. The next day he was in a busy, active delirium, but not so violent, with almost complete insomnia. He took nourishment. His temperature ranged from 100° to 101.4°; the pulse was about 120. It was noticed that his pupils were a little contracted, and that the left scarcely responded at all to light. He was seen frequently during the following night and was always awake. The next morning he was found much more feeble, and stimulants were ordered. Most of the time he was delirious, but had short lucid intervals, in which he recognized friends. He sank quite rapidly during the day; the delirium became low and muttering, and he died unconscious at 11 P.M., five days after admission.

Autopsy.—Fourteen hours after death. The head and spine only were examined. The skull was normal. The brain itself showed intense congestion, but nothing else pathological.

THREE CASES OF FRACTURE OF THE SPINE, WITH AUTOPSIES.

The spine was sawn through vertically. Fig. 190 shows the outside of the left half. Fig. 191 the inside view of the right half of the cervical and upper dorsal vertebræ.

Fig. 190.

Fig. 191.

The anterior and posterior common ligaments opposite the sixth and seventh cervical vertebræ were found to have been stripped from the bones, but not ruptured. There was a small extravasation of blood between the posterior ligament and the bone. The sixth vertebra was separated from the intervertebral disc, and with the vertebræ above it displaced slightly forwards, but not enough to have produced much pressure upon the cord. (See Fig. 191.) The vertebral arch and processes of the sixth cervical vertebra were the only ones affected. The lesion here was different upon the two sides, corresponding quite closely to the difference in the paralysis upon the two sides noticed during life. On the right side the pedicle was broken off close to the body, and the posterior part of the transverse process was separated from the anterior part. There was here very little displacement. On the left side the lesion was much more extensive. The pedicle was broken, as on the right side, but a little further back. The inferior articular process was displaced in front of the superior articular process of the seventh cervical. The transverse process was broken into several pieces, and so displaced as to produce very great pressure, and very likely some laceration of the sixth cervical nerve, just beyond its exit from the canal. The spinous process was also fractured close to its origin, but was not especially displaced. The gross appearance of the cord did not show any laceration. No evidence of spinal meningitis was seen, except to a very slight extent at the point of inquiry. The interpretation of the symptoms and lesions seems to me to be as follows:

The cause of death was delirium tremens, the exciting cause of which was the injury.

The paralysis in the upper extremities was principally of peripheral origin, produced by injury to the nerves in the intervertebral foramina, and just after they had passed through them. Some pressure upon the superficial fibres of the cord was probably caused by the slight extravasation and bony displacement. So far as the bony lesion was concerned there seems to be no reason why the patient should not have recovered from all except the paralysis of the left arm.

The autopsy in this case may furnish a clue to those by no means rare cases, in which localized paralyses are seen, but in which recovery takes place.

[193]

DOWNWARD DISPLACEMENT OF THE TRANSVERSE COLON.

THREE CASES, WITH AUTOPSIES.

CHARLES HERMON THOMAS, M.D.,

Surgeon to Philadelphia Hospital.

A deformity of the transverse colon, consisting in the elongation of that portion of the large intestine and its displacement downward in the form of a loop or festoon, has been observed by me in three instances in private practice. Autopsies were had in them all. In the first the most dependent portion of the gut was found midway between the umbilicus and the pubic symphysis; in the second it was deeply impacted in the cavity of the pelvis; and in the third it reached the level of the umbilicus.

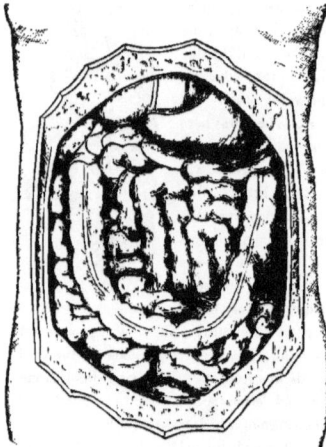 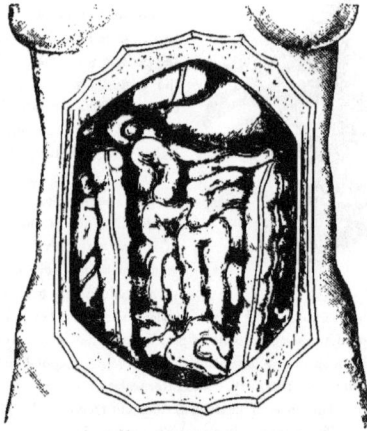

FIG. 192.   FIG. 193.

A positive diagnosis was not made in any of the cases, although in two of them the striking •
clinical conditions present were studied with special care in association with experienced and highly

[194]

skilled observers. In the second in order of occurrence, the relationship between it and the preceding one suddenly occurred to my mind, and was communicated to the operator while on our way to make the *post-mortem* examination. In the third case the actual condition present was strongly suspected before death. So that in both of these latter, special precaution was used at the autopsies to avoid disturbing the relative position of the abdominal viscera until their location had been accurately determined.

The lesion here described seems to be of rare occurrence. Thus far I have failed to discover a single recorded case; and not until this paper was nearly completed was I able to find any published reference to the condition, however vague. Several months ago I asked the assistance of Dr. Formad, who informed me that in a series of autopsies, numbering over 2,000, which he had made, he had not observed an instance of like character. He has also kindly sent me the following note:

"University of Pennsylvania, Dec. 15, 1882.

"*Dear Dr. Thomas :*

"* * * I looked very thoroughly through the literature of intestinal lesions, but did not meet any record of misplacement of the transverse colon.

"Very truly yours,

"H. F. Formad."

Case I.—Male, *act.* 80 years, a retired gentleman, came under my care August, 1874, as a patient of Dr. J. J. Levick, who had placed his practice in my charge during his vacation, and who informs me that there was no previous history of abdominal disease.

The symptoms present were extreme emaciation, feebleness, anorexia, and a profuse but fitful diarrhœa. The abdomen was retracted and somewhat tender upon pressure. There was no complaint of pain except at intervals of three or four hours when the diarrhœa had ceased for a time. Coincidently with the cessation of the diarrhœa a tumor about five inches long and two inches wide, of firm consistency, and visible on inspection, appeared beneath the thinned abdominal walls in a transverse position midway between the umbilicus and the symphysis pubis. The tumor persisted but an hour or so at a time, disappearing immediately upon the return of the diarrhœa. During the periods of continuance of the tumor the pain was so severe as to rapidly weaken the patient. This condition of alternate flux and painful tumefaction was repeated several times daily until death took place. During the attendance upon the case there were associated with me Dr. Albert H. Smith and a distinguished physician from another city—a near relative of the patient. With attention fully directed toward it, and after repeated observations, we were unable to frame a reasonable hypothesis as to the exact character and origin of the tumor. Death occurred September 12th, about three weeks from date of attack.

Autopsy. - In the presence of Dr. Levick and the relative mentioned, I made the abdominal section. To the former I am especially indebted for the specimen obtained, and which is still preserved.

Upon laying open the abdominal cavity the transverse colon was found to be greatly elongated and proportionately narrowed, the loculi being nearly obliterated, forming a loop open at the top similar to the letter U (Fig. 192), the most dependent portion occupying the position of and constituting

[195]

the tumor as above described, i. e. the horizontal portion of the loop rested upon the small intestines, midway between the umbilicus and the pubic symphysis.

II.—Female, æt. 54 years, a lady of delicate frame and refined habits of life, was under my charge for about ten months prior to her decease. During the greater portion of this period Dr. Jas. H. Hutchinson was associated with me in the attendance. Dr. Chas. K. Mills also saw her for me during my vacation. The patient had previously been attended by a homeopathic practitioner who had diagnosticated her condition as enlargement of the liver and stricture of the rectum. The latter supposed condition he had treated by the introduction of rectal bougies; this practice being afterward abandoned on account of the pain produced, and the lack of beneficial results.

Profound cerebrasthenia from other causes, with several months of delirium, and which finally led to a fatal result, served greatly to complicate the issues involved. The abdominal conditions which had been recognized from the beginning were thus either masked or placed entirely in abeyance during much of the time.

The more prominent symptoms recognized were (1) pain, referred chiefly to the region of the liver and extending both upward and downward, which pain was aggravated by walking, and was described as of a dragging, tearing character, and which had existed for four years or more. It was very much relieved by the recumbent posture, and after some months spent mostly in bed almost entirely vanished.

(2) Obstinate constipation with indications of obstruction, even a liquid passage being voided with difficulty. The capacity of the rectum to retain enemata also was diminished to two ounces.

(3) Two solid tumors elongated in form and of the consistency of solid fæces were discovered, located one on each side of the abdomen, and evidently just beneath the parietal structures. They were vertical in position, and about eight inches distant from and so parallel to each other, and were traced from the border of the ribs to within about two inches of the pelvic brim. This condition was observed but a few times, and at considerable intervals; at other times it was absent. The hypothesis was adopted that these masses were the ascending and descending colon, respectively, in a state of fæcal impaction.

Death occurred March 30, 1883, supervening upon a severe mental shock. An autopsy was made by Dr. Wm. M. Gray two days later, Dr. Hutchinson and myself being present. To quote from Dr. Gray's notes: " Upon opening the abdomen found complete prolapse of the transverse colon (Fig. 193). It was carried beneath the pubis and rested on the bladder. The large intestine was much narrowed, and was filled throughout with hard nodulated fæces ; the meso-colon was absent and the omentum, which was free from fat, was extremely atrophied : the rectum was normal, showing no evidence of stricture; the liver was of normal size, but upon microscopic examination showed marked cirrhosis."

Thus, that which had appeared to be the ascending colon proved to be the descending limb of the displaced transverse colon; and that which had seemed to be the descending colon was shown to be the ascending limb of the same malformation.

The pain which had previously been felt in the region of the liver and which had been relieved by recumbency had manifestly been caused by the sharp flexure of the colon contiguous to it ; and the rectal obstruction by the crowded condition of the pelvis produced by the invading loop of large intestine.

DOWNWARD DISPLACEMENT OF THE TRANSVERSE COLON.

Case III.—Male, æt. 30 years, a tailor's cutter, under attendance nine days prior to decease. He was the subject of advanced Bright's disease, with "hyaline, epithelial and granular tube casts, mucous cells, compound granule cells, and free oil globules." He also complained of severe pain in the abdomen to the right of and slightly above the level of the umbilicus. Upon inspection and palpation of the part no enlargement or induration was discovered; but light percussion developed an intensely tympanitic sound confined to the region described. Misplacement of the transverse colon was suspected, and the region kept under observation for any evidences of fæcal impaction which might, but which did not, present. Death occurred suddenly March 19, 1883. *Autopsy* two days later, Dr. Wm. M. Gray operator, Dr. Wm. H. Burke and myself being present.

The following notes were made by Dr. Burke. * * * "Body rather emaciated, and showing signs of commencing decomposition. On opening abdomen absence of fat noted, omentum normal. Peritoneum showing traces of lymph and pus, in the pelvic region especially, but no general inflammation. "Transverse colon empty, distended with gas, and has a sharp flexure at its center, bending obliquely downward and toward the right, to the level of the umbilicus, thence sharply upward to its normal position at the hepatic flexure. Meso-colon intact and apparently normal except in length. No sign of fæcal obstruction at the point of flexure. Both kidneys scirrhotic; capsule adherent, and secreting structure destroyed."

Evidently the heightened tympany localized near the umbilicus, which had been previously recognized and ascribed to the presence there of a portion of the transverse colon misplaced, had in reality been so caused.

No adhesions of the displaced parts were found in any of the cases cited. The intestinal fault was probably not the cause of death in any of them. Taking them together it will be seen that clinical conditions and *post-mortem* appearances agree in at least one important particular, viz., the location of the displaced intestine in contact with the anterior abdominal wall and below its normal site.

The anatomical relations of the ascending and descending colon respectively, it should be borne in mind, are in contact with the *posterior* abdominal walls, behind the small intestines, being bound down by reflections of the peritoneum. It therefore seems a practical impossibility for these portions of the colon to become spontaneously misplaced anteriorly. But of the transverse colon which is normally in contact with the *anterior* wall, in front of the small intestines, where it is suspended by the longer transverse meso-colon, these cases show that its displacement downward is of repeated occurrence.

Conclusions.—(1) Displacement of the transverse colon downward within the abdomen may be to any degree, partial or complete.

(2) Such displacement will present as solid tumor if the bowel be in a state of fæcal impaction, or as a limited area of heightened resonance if the bowel be distended with gas; but in either case the displaced part is to be found *in contact with the anterior abdominal wall.*

(3) The occurrence of intra-abdominal tumor situated below the normal site of the transverse colon, and having the same general configuration as the colon, such tumor being of a certain consistency, and presenting evidences of being in contact with the anterior abdominal wall; or, the occurrence of areas of special tympany with like outlines and similarly located, constitute diagnostic signs strongly indicative of downward displacement of the transverse colon.

A CASE OF SYPHILITIC DACTYLITIS.

BY HERBERT G. LYTTLE, M.D.,

Associate Professor of Genito-Urinary and Venereal Diseases in the New York Post-Graduate Medical School.

Edward E., two years and three months old, was brought by its mother (a mulatto) to my class at the New York Dispensary, in August, 1880, with the following history. He had been perfectly healthy until January, 1880, when he commenced to lose his appetite and flesh. His mother stated positively that he had never had snuffles or any rash upon the body. In March she first noticed a swelling of the ring finger of the left hand. Shortly after this the index finger of right hand commenced to enlarge, following this the index of left, and middle finger of right hand. In the latter part of April the second toe of left foot began to swell, and in May the first metatarsal of right foot became involved. These swellings gradually grew larger, attaining their maximum size in from four to six weeks, without pain or constitutional disturbance. He was under treatment from May, 1880, until he came under my observation in August, but without much benefit.

FIG. 194.   FIG. 195.

At this time he was small for his age, but fairly nourished. The following measurements were taken: First phalanx of left ring finger, 2⅝ inches in circumference; first phalanx of right index, 2⅜ inches; second phalanx of left index 1⅞ inches; second phalanx of right middle finger 2 inches in circumference. The second phalanx of second toe of left foot 1⅝ inches. The circumference of the right foot over the enlargement of the metatarsal bone is 7⅞ inches.

[198]

A CASE OF SYPHILITIC DACTYLITIS.

The swelling of the first phalanges was oval in shape, normal in color, smooth, tense, and the skin freely movable. The swelling of the second phalanx in each case was more irregular: the skin over a part of the swelling inflamed, of a dark coppery color, and at one point there was a small ulcerated spot discharging a thin grumous material. The child was placed upon the so-called mixed treatment, and the swellings dressed with plaster spread with mercurial ointment. The medicine did not agree with his stomach, and after two weeks treatment was changed to mercurial inunctions, with the iodide of potassium internally in increasing doses, beginning with three grains twice daily. At the end of two months there was a slight improvement in the ulcerations, and but little change in the size of the swellings. After this time the mother attended very irregularly at the dispensary for a few weeks, and then not at all until the next April. During this time he was treated for rheumatism. His general condition was very bad. The first phalanges of ring and index fingers had broken down and were ulcerating. The mother stated that small pieces of bone had from time to time come away in the discharge from all of the second phalanges, and they have diminished in size. He was given syr. ferri iodidi, ol. morrhuæ and mercurial inunctions. A decided improvement in his general condition followed this treatment, and he was again given the mixed treatment. In July, 1881, the second phalanges had nearly reached their normal size. Small pieces of bone had been discharging from the first phalanges, and the ulcerations had nearly healed. After this date the mother again became careless in her attendance, and the child did not receive thorough treatment. In September the second phalanges were entirely healed. There was a deformity in each, but greatest in left index, which was twisted to the left, and there was a false joint in both middle and index fingers, due to loss of bone. The first phalanges were not so much improved. The second phalanx of toe and meta-tarsal of right foot were improving. In November the child's condition was very good, the only trouble being the ulcerations of first phalanges. The mother again absented herself for a period of about four months, the boy not receiving any treatment in the mean time. In January, 1882, she noticed that the left arm at the elbow was swollen, and one month later the third metacarpal of right hand became affected. The first phalanges were in about the same condition as at last visit; swelling at elbow, seven inches in circumference. He was under treatment for about two months, with slight improvement. I then lost sight of him until January of this year, he being without treatment since the previous May. In July, 1882, the first metacarpal bone of left hand commenced to enlarge, and in November a swelling appeared at the wrist.

In January, 1883, his general health was quite good. The first phalanges were in the same condition as when seen last May. There was a depression over third metacarpal bone, at the bottom of which was a small opening, leading into a sinus one-quarter of an inch in depth. Small pieces of bone had been discharged from this opening. The metatarsal of right foot was still enlarged, the skin inflamed, with a pin-hole ulceration and thin, watery discharge. Circumference of foot, 7¾ inches; left foot, 5¼ inches. The swelling at left elbow measured 8¼ inches; right arm, 5 inches. The muscles of the left arm had wasted away from lack of use, as he was unable to either flex or extend the forearm more than half way. The osseous swelling involved the olecranon process of the ulna and the head of the radius. There was no redness and no pain except on pressure. The first meta-carpal of left hand was more swollen, inflamed, and ulcerating. The swelling at the wrist involved both the radius and ulna. Just below the styloid process of the ulna was an inflamed swelling, which gave fluctuation and resembled a gumma. Mixed treatment.

In February the swelling of wrist broke down and discharged freely a thick matter. In March photographs of the hands were taken. The right shows very well the first phalanx still enlarged and shortened; the loss of bone and resulting deformity of second phalanx; the depression over third metacarpal, and the large ulcerating surface at wrist. The cut of the left hand gives a very poor idea of its condition. He was kept steadily under treatment, but in April the swelling at elbow became inflamed and more swollen, causing quite severe pain and constitutional disturbance. He lost flesh at this time, and looked very bad. In May the swelling at elbow broke and discharged freely, followed by great improvement in his general condition. During June and July he improved very fast, except first metacarpal of left hand and first phalanges, which became inflamed and ulcerated. In September wrist and elbow had improved very much. He had been under treatment since January, and was in better condition than I had ever seen him.

The swellings which appeared last have been more thoroughly and continuously under treatment than those appearing first, and show much better results.

There are several points of interest in this case. Is it a case of hereditary or acquired syphilis? I have been unable to obtain any history of syphilis in either parent, the mother denying positively every symptom of syphilis, or having been under any treatment. She has had no miscarriages, but has borne two *healthy* children since the birth of this first one. Their ages are three and two years. I have watched both of them very closely, and they have never developed a symptom of syphilis up to this date. It is possible that the father of the first child is not the father of the subsequent children. In favor of the acquired origin of the disease might be noted the late appearance of the lesions, the child enjoying good health until it was a year and nine months old. There is no history of snuffles or any rash upon the body up to the time of the appearance of the bone swellings, but the mother is not very observing, and the child may have had them.

The bone lesions in acquired syphilis, when both forearms are involved, however, are usually symmetrical. In this case they are not so, the lower part of the right radius and ulna and the upper part of the left being affected.

It is stated by Taylor, in his monograph on "Bone Syphilis in Children," that there are no recorded cases of hereditary dactylitis, where the second phalanges of the fingers or phalanges of the toes are involved, in infants. He does not state at what age infancy ends, but according to the general acceptation of the term it extends to two and a half years. In this case these lesions were developed before the child was two years old ; but, as Dr. Taylor's cases were nearly all under *one year,* he might not consider this case as coming under the head of infancy.

---

* Two to three months. One fourteen months. His case of acquired syphilis age not given.

www.ingramcontent.com/pod-product-compliance
Lightning Source LLC
Chambersburg PA
CBHW021517210326
41599CB00012B/1286